Brain Bytes

Additional Acclaim

"Where should you go to find the answer to some quirky question about the brain? Right here! *Brain Bytes* has something for everyone! For the curious, browse at will. For the aficionado, find definitions, history, and clear functional explanations. Chudler and Johnson debunk myths, demystify terminology, and clarify how the brain works to control our bodies and make our minds. They relate experimental results in non-technical prose, candidly pointing out how correlative, not causative, evidence came to be represented as common misconceptions."

—**Janet M. Dubinsky, PhD**, Professor, Department
of Neuroscience, University of Minnesota

"I loved this book! The information is conveyed in an easy-to-understand manner, but also delves more deeply into the analysis of what's happening on a neurological level for those who wish to know more. Educators will appreciate the question-first style, as it allows for easy reference to student questions. For educators, many of the chapters are great for jumpstarting creative neuroscience lessons, starting a fish bowl discussion or Socratic seminar, or for use as a writing prompt. What a great gift!"

—**Brigitte Tennis**, National Teacher's Hall
of Fame 2015, Headmistress & Eighth Grade
Teacher, Stella Schola Middle School

"*Brain Bytes* is a delight from start to finish: a page-turner I couldn't put down. The topics are inherently interesting, writing style engaging, and scholarship top notch. Even the appendices are of great value. I recommend this to anyone with an interest in brain and behavior, professional or novice."

—**J. Timothy Cannon, PhD,** Professor of
Psychology, University of Scranton

Brain Bytes

Quick Answers to Quirky Questions
About the Brain

Eric H. Chudler and Lise A. Johnson

Illustrated by Kelly S. Chudler

W. W. Norton & Company
Independent Publishers Since 1923
New York London

Copyright © 2017 by Eric Chudler and Lise A. Johnson
Illustrations copyright © 2017 by Kelly S. Chudler

For information about permission to reproduce selections from this
book, write to Permissions, W. W. Norton & Company, Inc.,
500 Fifth Avenue, New York, NY 10110

For information about special discounts for bulk purchases, please
contact W. W. Norton Special Sales at specialsales@wwnorton.com or
800-233-4830

Manufacturing by Edwards Brothers Malloy
Book design by Nick Caruso
Production manager: Christine Critelli

Library of Congress Cataloging-in-Publication Data

Names: Chudler, Eric H., author. | Johnson, Lise A., author.
Title: Brain bytes : quick answers to quirky questions about the brain /
Eric Chudler and Lise A. Johnson ; illustrations by Kelly S. Chudler.
Description: New York : W.W. Norton & Company, 2017. | Includes
bibliographical references.
Identifiers: LCCN 2016029072 | ISBN 9780393711448 (hardcover)
Subjects: LCSH: Brain--Miscellanea. | Neurosciences--Miscellanea.
Classification: LCC QP376 .C493 2017 | DDC 612.8/2--dc23 LC record
available at https://lccn.loc.gov/2016029072

W. W. Norton & Company, Inc.
500 Fifth Avenue, New York, N.Y. 10110
www.wwnorton.com

W. W. Norton & Company Ltd.
15 Carlisle Street, London W1D 3BS

1 2 3 4 5 6 7 8 9 0

Important Note: *Brain Bytes* is intended to provide general information
on the subject of health and well-being; it is not a substitute for medical
or psychological treatment and may not be relied upon for purposes
of diagnosing or treating any illness. Please seek out the care of
a professional healthcare provider if you are pregnant, nursing, or
experiencing symptoms of any potentially serious condition.

CONTENTS

ANCIENT NEUROSCIENCE

WHAT'S UNDER THE HOOD?

PEOPLE

INTELLIGENCE

MEMORY

SLEEP

THE SENSES AND PERCEPTION

DRUGS, VENOMS, AND ADDICTION

POPULAR CULTURE

TECHNOLOGY

MEDICINE

BRAIN HEALTH

APPENDICES

ACKNOWLEDGMENTS

We thank Steve Johnson for his careful reading of this book and for his many helpful comments and suggestions.

INTRODUCTION

Neuroscience is everywhere. From magazine covers to Hollywood blockbusters, the brain is front and center. This popular media exposure has inspired many questions from people who wonder just what is going on in the three pounds of tissue between their ears. Answers to some of these questions have been provided by neuroscientists, whereas answers to other questions remain unclear. In this book we answer some unusual questions about the structure and function of the brain. We encourage you to verify our responses with other resources. We hope you will be surprised by our answers and, most important, we hope to motivate you to ask more questions. There is still much to learn.

Brain Bytes

ANCIENT NEUROSCIENCE

? **?** **?** **?** **?**

Did people always believe that the brain was important?

A

The brain was not always held in high esteem. For example, while preparing mummies, the ancient Egyptians preserved the heart, liver, stomach, intestines, kidneys, and lungs of the deceased, but they scooped out the brain through the nose and threw it away. The heart, not the brain, was thought to be responsible for thinking, sensing, and feeling. The ancient Egyptians, however, are credited for the oldest written record using the word *brain* in the Edwin Smith Surgical Papyrus.

Greek philosopher and physician Hippocrates (460–377 BCE) asserted that the brain was responsible for intelligence and emotion. This line of thinking differed from that of the Egyptians and was likely the result of Hippocrates's observations of people who suffered from brain damage. Hippocrates also attributed epilepsy to a disorder of the brain, not to a supernatural source.

Although Hippocratic views about the brain were adopted by many later scholars, Greek philosopher Aristotle (384–322 BCE) held on to the belief that the heart was the seat of sensation and movement. According to Aristotle, the brain served only to cool the blood. Greek physician Galen (129 CE–?) was well aware of the views of Hippocrates and Aristotle. Galen dissected the brains and nerves of many different animals, but not those of humans. Nevertheless, his experiments with animals and his observations of people after they suffered head injuries led him to reject Aristotle's view about the importance of the heart and agree with the Hippocratic view that the brain was the location of the mind.

Galen was wrong about his many descriptions of human neuroanatomy because he had only nonhuman brains to study. Regardless of these errors, his teachings about the nervous system persisted for hundreds of years.

For more information about the history of neuroscience, turn to Appendix 6 for a list of many milestones in brain research.

Reference

Finger, S., *Origins of Neuroscience* (New York: Oxford University Press, 1994).

Did people really believe that the bumps on a person's head would say something about a person's intelligence and personality?

A

Reading a person's personality, character, and intelligence from the bumps on their head was very popular in the 1800s. This method, called phrenology, was developed by German physician Franz Joseph Gall (1758–1828) and popularized in the United States by Johann Spurzheim (1776–1832).

Phrenology was built on the idea that distinct personality characteristics are located in specific areas of the brain, and bumps or dents on the skull reflected the strength or weakness of these brain areas. Gall used the skulls of prominent writers, scholars, and politicians as well as criminals and people with cognitive problems to construct his model because he thought specific traits and their cranial roadmaps would be more apparent in these individuals. His original maps had 27 different personality characteristics including pride, cunning, wisdom, poetic talent, memory for people, memory for words, and sense of numbers. Spurzheim later added eight more characteristics to the map.

By feeling a person's skull, phrenologists believed they could deduce that person's character and personality. The reading could be used to help people improve themselves. Of course, not everyone was impressed with phrenology. Critics pointed out that injuries to the brain did not result in behavioral changes associated with the skull maps. Other scientists, such as French physiologist Jean Pierre Flourens (1794–1857), argued against the notion that the cerebral cortex had specific functions at all. Rather, he promoted the idea that all areas of the cerebral cortex functioned equally. Flourens was right about phrenology, but wrong about the cortex.

The idea of phrenology seems strange to us now, and indeed, it was eventually completely debunked by the scientific community. But it did promote the scientifically correct idea that different parts of the brain were specialized for different functions.

? ? **?** **?** ?
Q

Has a neuroscientist ever won a Nobel Prize?

A

Neuroscientists have received their share of Nobel Prizes in Physiology or Medicine. In fact, 60 neuroscientists have been awarded (individually or shared) a Nobel Prize since the first prize was given in 1901. The first neuroscientists to win were Camillo Golgi (1843–1926) and Santiago Ramon y Cajal (1852–1934). Although they were scientific rivals because of their differing opinions about how neurons formed connections, Golgi and Cajal shared the 1906 Nobel Prize in Physiology or Medicine for their work describing the structure of the nervous system, although they didn't share it graciously. Cajal's theory about a gap between neurons was eventually proven to be (mostly) correct.

Later neuroscientists were recognized with Nobel Prizes for their studies about neurological disorders, hearing, vision, smell, neurochemistry, brain imaging, and neurotransmission. The 1949 prize given to Antonio Egas Moniz (1874–1955) was very controversial because of its focus on the development of the lobotomy to treat mental illness. In retrospect the award to Moniz seems like a poor choice, but the prize has not been retracted. The most recent neuroscientists to win the Nobel Prize are John O'Keefe (born, 1939), Edvard I. Moser (born, 1962), and May-Britt Moser (born, 1963) who won in 2014 for their work on the spatial navigation system in the brain.

Appendix 5 has a list of all neuroscientists who have won the Nobel Prize in Physiology or Medicine.

How did scientists learn that nerve cells used chemical messages?

A

Some people, such as German scientist Emil Du Bois-Reymond (1818–1896), are ahead of their time. In the late 1800s, Du Bois-Reymond proposed that nerve cells could communicate with muscles using chemicals. At the time, no one took notice of his theory; researchers were still debating whether nerve cells made electrical or chemical connections.

It was not until the early 1900s, when scientists started to test the effects of drugs on the autonomic nervous system, that Du Bois-Reymond's idea was confirmed. For example, English pharmacologist Henry Hallett Dale (1875–1968) experimented with ergot, a fungus that grows on rye plants. He found that ergot could block the effects of adrenaline and that other drugs could reproduce the effects of adrenaline. Dale was hesitant to call these drugs chemical messengers (neurotransmitters) because he did not know if the body contained these chemicals naturally.

Enter German pharmacologist Otto Loewi (1873–1961). He knew of Dale's work and conducted his own research into how drugs affected the function of glands and internal organs. In 1921, Loewi went to sleep one night and dreamed of an experiment that would revolutionize our understanding of the nervous system. When he woke up in the middle of the night, he wrote down the important experiment and then went back to sleep. The next morning when he arose, he could not read his own writing and could not remember the contents of his dream.

As luck would have it, Loewi had the same dream of the

experiment the next night. This time when he woke from his dream, he did not take any chances. He went straight to his laboratory to do the experiment. In this experiment, Loewi took one frog heart that was still connected to the vagus nerve. This heart was put into a container with fluid. The container with the heart was connected to a second container that held another frog heart such that the fluid from the first container could flow into the second container. When Loewi electrically stimulated the vagus nerve of the first heart, the first heart slowed down. Interestingly, after a short time, the rate of the second heart that received fluid from the container holding the first heart also slowed down. Loewi's conclusion was that the vagus nerve released a chemical into the first container that then affected the heart in the second container. He named this chemical *Vagusstoff*. We now know it as the neurotransmitter acetylcholine.

This is a great example of sleep facilitating creativity and problem solving, although this is an admittedly extreme case. For their investigations into the chemical transmission of nerve impulses, Dale and Loewi shared the 1936 Nobel Prize for Physiology or Medicine.

Reference

Loewi, O., *From the Workshop of Discoveries* (Lawrence: University of Kansas Press, 1953).

What is trepanation?

A

In basic terms, trepanation is putting a hole in someone's head. More precisely, it is the removal of a portion of the skull, which exposes, but does not puncture, the membrane covering the surface of the brain.

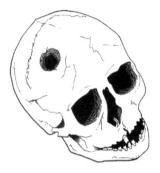

Trepanning (or trephination) is the oldest known surgical procedure. Archeologists have uncovered skulls at least 10,000 years old that show evidence of trepanation. The practice was quite widespread: ancient specimens have been recovered from all over the world including Europe, Asia, Africa, South America, Tahiti, New Guinea, and New Zealand. Amazingly, most of the people who received trepanation survived the procedure, as evidenced by the healing of their skulls. In most cases it isn't clear why the procedure was performed, although some people speculate that practitioners may have believed that opening the skull would release evil spirits responsible for an illness or disease. Trepanning could relieve intracranial pressure from a traumatically induced hematoma, which would have been therapeutic. It may also have been used to treat headaches, epilepsy, and mental illness—ailments for which the therapeutic value of the treatment is less clear.

Although this procedure may seem like an interesting historical note, there is a small but vocal contingent of people today who advocate trepanation as a way to achieve an enhanced state of consciousness. This theory was developed in the 1960s by Bart Huges, a Dutch medical student and recreational user of psychoactive drugs. Under the influence of these drugs, Huges became convinced that drilling a hole in his head would allow blood and cerebrospinal fluid to move more freely around his brain, mimicking the state of an infant's brain prior to the fusion of the skull bones. He tested this theory by drilling a hole in his own forehead with a foot-operated dental drill. He considered the procedure a tremendous success and subsequently convinced others to follow his lead, most notably Amanda Feilding, who filmed her self-trepanation in 1970. This film (*Heartbeat in the Brain*) alternates shots of the surgery with still shots of Feilding's pet pigeon. You can find the film online (but be warned, it is graphic). Feilding is still a vocal advocate of trepanation and has established the Beckley Foundation to fund (among other things) scientific research on consciousness. The attitude of scientists and physicians, however, remains overwhelmingly skeptical about the benefits of recreational trepanation.

References

Gump, W., Modern induced skull deformity in adults, *Neurosurgical Focus*, 29:E4, 2010.

Parapia, L.A., Trepanning or trephines: a history of bone marrow biopsy, *British Journal of Haematology*, 139:14–19, 2007.

WHAT'S UNDER THE HOOD?

? **?** **?** **?** **?**
Q

How did parts of the brain get their strange names?

A

Scientific terminology is often difficult to understand. Science is filled with unusual words and phrases, and brain research is no exception. Amygdala, hippocampus, corpus callosum—why do these brain structures have these names, and what do they mean?

Do these and other names of brain areas sound like Greek and Latin to you? They should! Most terminology used in neuroscience has its roots in Greek and Latin. Some words originated with ancient Greek philosophers, physicians, and scientists, such as Hippocrates, Aristotle, Herophilos, and Galen. Other terms entered the medical and scientific vocabulary more recently as new discoveries were made and diseases and disorders were named after the people who first described them. Some terms are hybrids of Greek and Latin such as the word *neuroscience* from the Greek word *neuron* (nerve) and the Latin word *scientia* (science). Not all neuroscience words are so easy to decipher, but each attempts to provide a clear description of the place where it is located or the idea it is meant to define.

Although some words are just made up, most have a logic to their construction. Robert Fortuine provides an excellent framework to understand how medical terms are named, and his design can be applied to understand words used by neuroscientists:

> **LOCATION:** structure named based on its location in the body. For example, the lingual nerve is located in the tongue. "Lingual" comes from the

Latin word *lingua* meaning tongue. The hormone called adrenaline is derived from a word meaning "near the kidney."

FUNCTION: structure named based on what it does. The abducens nerve, for example, is from a Latin word meaning "drawing away" because this nerve controls an eye muscle responsible for moving the eye horizontally.

RESEMBLANCE: structure named based on what it resembles. Resemblance is perhaps the most common method for naming neuroanatomical structures. Examples include arachnoid ("spider web-like"), hippocampus ("seahorse" or "sea monster"), amygdala ("almond"), pineal ("pine cone"), and cauda equina ("horse's tail").

CHARACTERISTICS: structure named based on its shape, size, texture, number, or color. Melanin ("black"), lenticular ("lens-like"), and ceruleus ("blue") are all neuroscientific words based on their characteristics.

EPONYM: diseases, instruments, or structures named after the people who discovered, invented, or described them. Parkinson's disease, for example, was named after James Parkinson (1755-1824), the English physician who described the neurodegenerative disease in manuscript titled "An Essay on the Shaking Palsy" published in 1817. Another neurodegenerative disease, Alzheimer's disease, was named after Alois Alzheimer (1864-1915), who described this disorder in 1906. Betz cells, found in the cerebral cortex, were named after Vladimir Alekseyevich Betz (1834–1894); Purkinje cells, located in the cerebellum, were named after Jan Purkinje (1787–1869). Many surgical instruments, such as the Penfield elevator (Wilder Penfield, 1891-1976) and Dandy forceps (Walter Dandy, 1886-1946) are named after their inventors.

Appendix 1 lists of the Greek and Latin origins of many words in brain research and Appendix 2 lists eponyms in neuroscience.

References

Duque-Parra, J. E., Llano-Idárraga, J.O., and Duque-Parra, C.A., Reflections on eponyms in neuroscience terminology, *Anatomical Record Part B: The New Anatomist*, 289B:219–24, 2006.

Fortuine, R., *The Words of Medicine. Sources, Meanings, and Delights* (Springfield: Charles C. Thomas Publisher, 2001).

Koehler, P.J., Bruyn, G.W. and Pearce, J.M.S., *Neurological Eponyms*, (New York: Oxford University Press, 2000).

? ? ? ? ?

Do all animals have a brain and nervous system?

A

Virtually all animals have some sort of nervous system, but most do not have a spinal cord. Those without a spinal cord are the *invertebrates* (animals without backbones). This classification includes insects, spiders, worms, mollusks, sea stars and jellyfish. Approximately 80 percent of the animal biomass, more than 95 percent of all animal species, and 99 percent of the Earth's biodiversity are invertebrates. No matter how you look at it, humans belong in the minority.

Instead of a spinal cord, some invertebrates, such as jellyfish, hydra, and anemones, have a nerve net distributed throughout their body. Other invertebrates such as grasshoppers and lobsters have a nerve cord that runs down the underside of their body. The nerve cord may be connected to a centralized collection of neurons (a brain) in the head of some invertebrates, but other invertebrates (such as jellyfish) do not have a brain.

The sponge is an example of one of the few multicellular animals without a nervous system at all: no brain, no spinal cord, no neurons. The sponge does not need a nervous system because its structure allows water to bring in oxygen and nutrients and eliminate waste. A sponge doesn't really need to *do* much, so it doesn't really need a brain. Brains are very expensive in terms of calorie consumption—far too expensive to have on board for recreational purposes. So, if an animal doesn't need a brain, it's best not to have one.

It may seem that humans have little in common with the spineless creatures that so dramatically outnumber us. However, where nervous systems are concerned, we have

quite a bit in common, and we can learn a lot about ourselves by studying them. For example, the marine invertebrate called the sea hare (*Aplysia*) has played an important role in neuroscience. With a central nervous system containing approximately 20,000 neurons, *Aplysia* have been used to investigate sensory and motor physiology and behavior. In 2000, Eric Kandel (1929–) was awarded the Nobel Prize in Physiology or Medicine for his work studying the molecular and cellular mechanisms of learning and memory, primarily using *Aplysia* as his model animal.

References

Borrell, B., One-fifth of invertebrate species at risk of extinction, *Nature*, September 3, 2012; doi:10.1038/nature.2012.11341.

Lewbart, G.A., *Invertebrate Medicine* (Ames, IA: Blackwell Publishing, 2006).

How big is the human brain?

A

Here are the facts: the average adult human brain weighs 1.3–1.5 kg (about 3 pounds, the weight of a cabbage), contains approximately 86 billion neurons and at least the same number of glial cells, and accounts for about 2 percent of the total body weight of a person. The average width, length, and height of the brain, respectively, are 140 mm, 167 mm, and 93 mm (about the size of two large fists). The composition of the brain is 77–78 percent water, 10–12 percent lipid (fat), 8 percent protein, 1 percent carbohydrate, 2 percent soluble organic substances, and 1 percent inorganic salts.

It takes several years for the brain to reach its full size. During development before birth, neurons are created at an incredible pace, with 250,000 neurons added every minute. Most of the neurons that the brain will ever have are present at birth, but the brain continues to grow after that. A newborn baby's brain weighs less than 400 g (about as much as a large orange), but by the age of two years, a baby's brain is

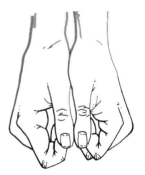
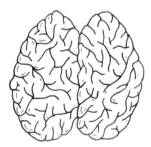

about 80 percent of its adult size. The increase in size is the result of new glial cells that continue to divide and multiply and the new connections the brain makes over time. Maximum brain weight occurs when a person is in his or her late teen years. The peak is brief: after age 25 the brain starts to atrophy, and brain weight declines.

In humans, the largest area of the brain by volume is the cerebral cortex. The cerebral cortex accounts for 77 percent of the total volume of the central nervous system. The many bumps (gyri) and grooves (sulci) work to expand the amount of cerebral cortex that can fit within the skull: the total surface area of the cerebral cortex is 2,500 cm² (about the size of a large pizza). The sheet of cerebral cortex, the layer that overlies other parts of the brain, ranges in thickness from 1.5 to 4.5 mm, which is not very thick. The cerebral cortex is divided into four lobes: the frontal lobe accounts for 41 percent of its total volume, the temporal lobe 22 percent of its volume, the parietal lobe 19 percent of its volume, and occipital lobe 18 percent of its volume. The next largest part of the central nervous system is the cerebellum, which makes up 10 percent of the total central nervous system volume. The diencephalon and midbrain each contribute 4 percent to the volume, and the hindbrain and spinal cord contribute another 2 percent each.

References

Kennedy, D.N., Lange, N., Makris, N., Bates, J., Meyer, J., and Caviness, V.S. Jr., Gyri of the human neocortex: an MRI-based analysis of volume and variance. *Cerebral Cortex*, 8:372–84, 1998.

Peters, A., and Jones, E.G., *Cerebral Cortex. Vol. 1, Cellular Components of the Cerebral Cortex* (New York: Plenum, 1984).

Does size matter, and do humans have the largest brains?

A

You might like to think humans have the largest brains in the animal kingdom, but we don't. In general, larger animals have larger brains. As you might guess based on head size, a cow has a larger brain than a cat and a cat has a larger brain than a mouse. One reason larger animals need a larger brain is to control bigger muscles and process more sensory information from the skin.

The adult human brain weighs about 1.4 kg (3 pounds). Several animals have brains larger than 1.4 kg. For example, elephants, bottlenose dolphins, pilot whales, humpback whales, killer whales, blue whales, and sperm whales all have brains larger than those of humans.

Do you think humans have the largest brains in proportion to our body size? Think again. Several animals have larger brain-to-body-weight ratios than humans. A 150-pound human with a 3-pound brain has a brain-weight-to-body-weight ratio of 1:50; the brain accounts for 2 percent of a person's total body weight. A hummingbird's brain accounts for 3.8 percent of its body weight.

An animal's brain size can also be compared to the brain size of what it is expected to be based on that of similar animals. If you know the actual weight of a brain and compare it to the expected brain weight of similar animals, you get the *encephalization quotient* (EQ). In general, brain weight is proportional to the two-thirds power of body weight. Using this formula it is possible to estimate how large a brain should be. Humans are at the top of the EQ scale with a value of 7.44.

Appendix 3 lists the weights of the brains from many different animals.

Reference

Rehkämper, G., Schuchmann, K.L., Schleicher, A., and Zilles, K., Encephalization in hummingbirds (Trochilidae), *Brain, Behavior and Evolution*, 37:85–91, 1991.

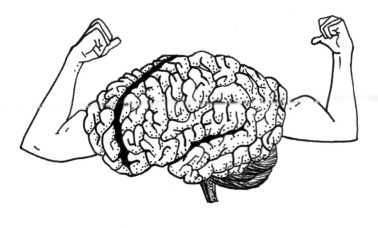

Is the brain a muscle?

A

You may have heard the saying: "Exercise your brain!" Does this mean you can strengthen your brain muscles? No, because the brain is not a muscle. Rather, the brain is composed of approximately 80 percent water, 10 percent fat, and 10 percent protein, salts, and inorganic matter. Except for smooth muscle cells lining the blood vessels of the brain, there is no other muscle tissue in the brain. "Exercising the brain" refers to giving your brain a mental workout, such as when you read, play chess, and solve crossword puzzles.

Why does the brain have so many folds?

A

The real question is why the *human* brain has so many folds, because not all brains are so wrinkled. Human brains are very convoluted, chimpanzee brains less so, monkey brains less still, and so on, until you get to the humble shrew, with a brain about as smooth as a cue ball. Shrews are also quite dim (as mammals go) and humans are quite smart, so you might think there is a relationship between the number of wrinkles on a brain and the intelligence of an animal. You would be right. That brings us to the reason for all of those folds: they allow more cerebral cortex to fit into the skull.

The cerebral cortex is the outer shell of the brain, the most recent evolutionary addition that is important for higher-level functions. Roughly speaking, more cerebral cortex equals more smart. The cerebral cortex is about the same thickness in all mammals (except for whales and dolphins, which have a relatively thin cortex), but the surface area (relative to body size) is much greater in more cognitively capable species. Unfolded, the cerebral cortex of the human brain is about two and a half square feet. A skull large enough to accommodate that much area if it were flat would be awkward to walk around with (not to mention unpleasant to give birth to). Cortical folding makes it possible to keep all those neurons in a tidier package.

How is the nervous system connected to the other systems in the body?

A

Your nervous system is connected directly or indirectly with all of the other systems of your body, including the circulatory, skeletal, muscular, reproductive, digestive, urinary, integumentary (skin, hair), lymphatic, endocrine, and respiratory systems.

Communication between the nervous system (brain, spinal cord, nerves) and other body systems is a two-way street with signals going to and from the brain and other parts of the body. A good example of this reciprocal communication is between the nervous system and the muscular system. To move your arms and legs, your brain must send signals down to the brain stem, where they cross to the other side of your body. The signals then continue down to the spinal cord, where they communicate with motor neurons. Motor neurons send signals through their axons to muscles telling them to contract. Sensory receptors in muscles send signals back into the nervous system with information about the state of the muscle. The brain uses this information from muscles to adjust the amount of muscle contraction. This regular talk between the nervous system and muscular system allows you to hold objects with the proper grip strength so you don't drop anything, yet you can hold objects tightly and not crush them.

The connections between the nervous system and other body systems are also two-way streets. For example, the brain controls heart rate (circulatory system), moves bones (skeletal system), controls the release of hormones (endocrine system), stimulates defense mechanisms (lymphatic system), regulates breathing rate (respiratory system), and controls feeding and drinking behavior (digestive system), mating behavior (reproductive system), and elimination (urinary system). The nervous system even controls muscles connected to hairs on the skin. In turn, body systems send sensory information back into the nervous system to provide feedback about internal organs.

How are the two halves of the brain connected?

A

The corpus callosum is a large band of approximately 200 million nerve fibers (axons) that connects the right and left hemispheres of the brain. This connection allows the transfer of information between the right and left sides and helps coordinate the activities of the brain.

Although most people have a corpus callosum, a rare condition called congenital agenesis of the corpus callosum occurs in which people are born with a partial or completely absent corpus callosum. The symptoms of this disorder can be mild or severe depending on other brain abnormalities associated with the condition. Some people without a corpus callosum have normal intelligence, but others have severe cognitive impairment, seizures, or movement problems.

Some people who have epileptic seizures that cannot be controlled with medication undergo a surgical procedure to cut the corpus callosum. For reasons that are still unknown, seizures experienced by people after the corpus callosum is cut are often reduced or eliminated. These "split-brain" patients can no longer transfer information between the right and left sides of the brain because the main pathway between the sides is eliminated. Intelligence is usually unchanged after split-brain surgery, and only special tests to restrict information to one side of the brain will reveal any difference in the behavior of these people after surgery. For example, if visual information is sent to only the right side of the brain, most people after split-brain surgery cannot name an object presented to them, nor will they say that they have seen an object at all, because language abilities in most people are located in the left hemisphere. However, if you ask this same person to draw the object with their left hand (the hand that is controlled by the right side of the brain), they can do it without any problem. This is bizarre to watch, and probably equally bizarre to experience. It demonstrates that while our conscious perception seems unified, in fact it is modular. Roger Sperry (1913–1994), who worked with split-brain patients, was awarded a Nobel Prize for his research in 1981.

Reference

Paul, L. K., Brown, W.S., Adolphs, R., Tyszka, J.M., Richards, L.J., Mukherjee, P., and Sherr, E.H., Agenesis of the corpus callosum: genetic, developmental and functional aspects of connectivity, *Nature Reviews Neuroscience*, 8:287–99 (2007).

? ? **?** ? ?

Q

How big is a nerve cell?

A

Nerve cells (neurons) come in many different shapes and sizes. A typical neuron has four main parts: dendrites, cell body, axon, and axon terminal. Dendrites are the main areas that have connections (synapses) with other neurons, although the other parts of a neuron can also support these connections. Special receptors on dendrites bind chemical messengers (neurotransmitters) that results in an electrical signal that is sent to the cell body.

The cell body contains organelles such as the nucleus, ribosomes, and mitochondria that provide the neuron with its genetic material, synthesize proteins, and produce energy. The diameter of a neuron's cell body ranges from 5 to 10 microns for small neurons called granule cells up to about 100 microns for neurons in the spinal cord that connect to muscles. For reference, 100 microns is equal to 0.1 millimeters or 0.004 inches, slightly smaller than the size of the period at the end of this sentence.

Neurons have a single axon that extends from the cell body and conducts electrical activity away from the cell body toward the axon terminal. Axons range in diameter from about 0.2 to 20 microns. Axons can be very short, for example, less than 1 millimeter for some neurons in the brain. Some axons, such as those stretching from the cell body of a neuron in the spinal cord to a muscle in the foot, are very long at 1 meter or more in length.

Is it true that there are more synapses in the brain than there are stars in the universe?

A

The human brain has 86–100 billion neurons. Each of these neurons can make hundreds or thousands of connections (synapses) with other neurons. The entire brain has 100 trillion to 1,000 trillion synapses, and the cerebral cortex alone has about 60 trillion synapses.

Even the highest estimate for the number of synapses pales in comparison to the number of stars in the universe. In our humble Milky Way galaxy alone, there are 100 billion (or 10^{11}) stars. With 10^{11} to 10^{12} galaxies in the universe, we are talking about a boatload (or spaceship-load) of stars. Doing the math with these numbers shows that there are approximately 10^{22} to 10^{24} stars in the entire universe. Although the number of synapses in the brain is big, the number of stars in the universe is much larger.

References

Daniels, P., Restak, R., Gura, T., and Stein, L., *Body: The Complete Human: How It Grows, How It Works, and How To Keep It Healthy and Strong* (Washington, DC: National Geographic Press, 2007).

European Space Agency, How many stars are there in the universe, http://www.esa.int/Our_Activities/Space_Science/How_many_stars_are_there_in_the_Universe, accessed December 15, 2015.

Do we get more neurons after we are born?

Your body has amazing regenerative powers. When you cut your skin, you make new skin cells to heal the injury. When you break a bone, your body makes new bone cells to heal the fracture. Unfortunately, new neurons do not seem to appear when there is damage to the brain. This is one reason it is so important to take care of the neurons you have. Your brain can rewire itself to compensate for the loss of some neurons, but it is better if you don't lose nerve cells in the first place.

Although new nerve cells are not created to repair damage, there are parts of the brain that make nerve cells even in adults. The hippocampus, an area important for learning and memory, is one such area. Neurogenesis (the creation of new neurons) presumably facilitates learning and memory, but it is unclear exactly how. Some evidence suggests that exercise increases the rate of hippocampal neurogenesis and improves learning. So get moving if you want new nerve cells.

Did dinosaurs such as the *Stegosaurus* have two brains?

A

Many older books claim that the *Stegosaurus* dinosaur needed a second brain in its spine to control movement of its massive legs and tail. Although *Stegosaurus* did have an enlarged spinal column near its hips, there is no evidence that the neural tissue there served as a second brain.

Most animals with limbs have enlargements in the spinal cord to control the muscles for arm and leg movement. These swellings make room for neurons that process sensory information coming from the skin of the limbs and for neurons that send information to move muscles. Much of the control for walking takes place in the spinal cord and does not need the brain. Near the same area as this enlargement in stegosaurus is a region called the glycogen body that stores glycogen. Although the function of the glycogen body is not understood, it is definitely not a brain.

? ? **?** ? **?**

Q

Can you move objects with brain "power?"

A

Sure! Electrical signals from your brain can be sent down to your spinal cord, and then neurons can send messages to tell muscles to move. Electrical signals from neurons can also be recorded by electrodes on the scalp, on top of the brain or inside the brain. These signals can be used to control robotic limbs, motors, or computers (see discussion of "brain–computer interfaces").

This may not be the type of brain power you were interested in. Can the power of thought somehow move through the air to cause an action? In other words, is telekinesis possible? Sorry to disappoint you! The weak electrical signals generated by the brain do not travel far from their source. To our knowledge, there are no definitive experiments that have successfully demonstrated telekinetic abilities.

Up until 2015, anyone who could show they had the ability to "move objects with their mind" could pick up a cool $1 million. To claim their money, people had to visit James Randi in his laboratory in San Francisco and show that they had these special abilities. Randi, who retired in 2015, had offered the million-dollar challenge for 19 years. His money stayed in his bank account.

Reference

James Randi Educational Foundation, http://web.randi.org/about-james-randi.html, accessed January 14, 2016.

Does the brain really use electricity to send messages?

A

Yes, the brain uses electricity to send messages, but it is created differently from the electricity that comes out of the wall socket (it's also much lower voltage). Neurons pass messages to each other electrochemically using a signal called the *action potential*. The electrical current powering your home is carried by electrons; in your brain, the electrical current is carried by ions.

Ions are electrically charged particles, and the body uses several different kinds. The ions responsible for sending electrical messages within neurons are sodium, potassium, calcium, and chloride. Because the membrane that surrounds a neuron is semi-permeable, some ions can pass through special channels, while others are blocked. The unequal distribution of ions inside and outside of a neuron sets up an electrical potential difference (a voltage).

When a neuron is not active, its inside is negative relative to the outside by about 70 mV (millivolts). In addition to this charge difference, there is a difference in the con-

centration of ions inside and outside the cell; there are relatively more sodium ions outside the neuron and more potassium ions inside the neuron. Remember, opposites attract. Positive ions want to go where it is more negative, and concentrated chemicals want to go where they are less concentrated.

The action potential, sometimes called a spike or an impulse, is caused by rapid opening and closing of different ion channels. To trigger an action potential, an incoming signal causes sodium channels to open, making the cell more positive (depolarized) as positively charge sodium ions rush into the neuron. When the depolarization reaches about -55 mV, the neuron generates an action potential. If the neuron does not reach this threshold, there will be no action potential.

If all systems are go, the first step in creating an action potential is the opening of more sodium ion channels near the axon hillock (the part of the neuron located near the cell body where the axon starts). When sodium channels open, sodium ions rush into the neuron because there are many more sodium ions outside the cell relative to the inside. This increase in internal sodium depolarizes the neuron dramatically—at the peak of the action potential the inside of the cell reaches about +30 mV. Meanwhile, potassium ion channels start to open. They take a little longer to get going, but when they do, potassium ions rush out of the cell because the inside of the cell is (relatively) crowded with potassium. This efflux repolarizes the neuron (makes it negative again). At about the same time the potassium ion channels open, sodium ion channels start to close. This helps the inside of the neuron go back toward -70 mV. Because potassium channels stay open for an extended period of time, the voltage of the inside actually overshoots the mark and goes a bit lower than the resting level. After the potassium ion channels close, the ion concentrations go back to their resting levels and the cell returns to -70 mV.

This takes a long time to explain, so you might think it takes a long time to happen. The whole action potential happens within a few milliseconds. However, it is quite slow compared to how quickly electricity travels through the metal wires in your walls.

How fast do messages travel through nerves?

Nerves are composed of axons from many different neurons. Some axons are very thin (less than 1 micron in diameter) and some axons are thicker (20 microns in diameter) and wrapped with insulating material called *myelin*. The thickness of a nerve fiber and whether it is insulated by a myelin sheath determine how fast an electrical signal travels down an axon. Thicker axons conduct action potentials more quickly than do thinner axons, and axons with myelin send action potentials more quickly than those without myelin.

Signals within an unmyelinated, thin axon travel between 0.5–2.0 m/sec (1.1–4.5 miles/hr; 1.8–7.2 km/hr). Information related to pain, itch, and temperature is carried in these small axons. In large-diameter axons insulated with myelin, signals race along at speeds up to 120 m/sec (268.4 miles/hr; 431.9 km/h). Large-diameter axons transmit sensory information about touch and muscle sensation.

You might have had an experience to show how different sensory signals travel at different rates in your nervous system. If you have bumped your knee or stubbed your toe, you may have perceived the first sensation you feel as one of touch or pressure. This nonpainful sensation is carried by fast-conducting, large, myelinated axons. After a short delay, you may have felt a painful sensation because the messages carried by slower conducting axons take a longer time to reach your brain.

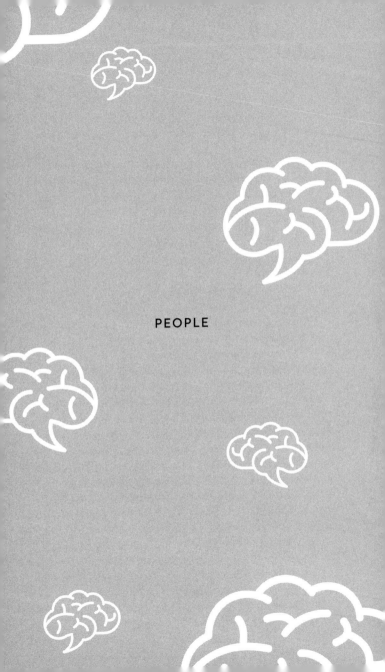

PEOPLE

What is the difference between a neuroscientist and a neurologist?

A

Neuroscience and neurology are both great professions, but people in these jobs have very different training and responsibilities. A *neuroscientist* is a person who studies the nervous system, and the word usually refers to a person who does research. A neuroscientist can study the structure, function, development, or chemistry of the nervous system or can investigate the effects of disease or damage to the nervous system. In contrast to a neuroscientist, a *neurologist* is a medical doctor who diagnoses and treats disorders of the human nervous system. Neurologists must understand the underlying causes of neurological disease, but they generally do not design experiments, collect data, or publish research results (although some do).

Most neuroscientists go to graduate school to earn a Ph.D. from a specialized department such as neuroscience, psychology, pharmacology, bioengineering, computer science, or biology. A graduate degree in one of these topics does not allow people to treat patients with neurological disorders or write prescriptions for medications. Physicians who have gone to medical school and specialize in neurology can examine patients, run tests to diagnose illness and recommend treatment, but if they do not conduct research, they are not neuroscientists.

Who was Phineas Gage?

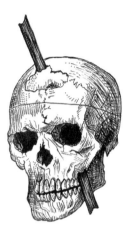

A

Depending on your perspective, Phineas Gage (1823–1860) was either very lucky or very unlucky. On the fateful day of September 13, 1848, in Cavendish, Vermont, Gage was working on the railroad when the tamping iron he was using to compact explosives exploded up through his cheek and rocketed through his brain and out the top of his skull. The 3-foot, 7-inch-long (13.5 pound) iron landed 25–30 yards away from him. Although Gage did not lose consciousness immediately after the accident, a large part of his left frontal lobe was destroyed. Gage secured his place in annals of neurology and neuroscience not only for surviving the accident but for his behavior after he recovered from his injuries.

Prior to the accident, Gage was described as a well-balanced, hard-working railroad foreman. Dr. John Harlow, the physician who treated Gage, mentions changes in Gage's behavior and personality after the accident, describing him as irreverent and impulsive. The changes caused friends to say that he was "no longer Gage" and resulted in his failure to get his job back with the railroad. Eventually, Gage found work driving horse coaches, first in New England and later in Chile.

Neuroscientist and historian Malcolm Macmillan believes that the behavioral problems were only temporary and that Gage was able to recover considerably. Over time, the brain could have rewired itself to compensate for the damage done by the injury.

Although the exact nature of Gage's injuries and subsequent recovery are not completely known, the case provided scientists with some of the first evidence of the specific functions localized to the frontal lobes.

Reference

Macmillan, M., Phineas Gage Information Page, November 2, 2012, accessed January 17, 2016, https://www.uakron.edu/gage/.

Who was H.M.?

A

A person stuck in the present and the past—unable to form new memories—as if time had stopped. This was the life of Henry Gustav Molaison (known by his initials, H.M.) after he recovered from brain surgery in 1953 at the age of 27.

In 1935 on his way to school, Molaison was struck by a bicycle and knocked to the ground, where he hit his head and lost consciousness for five minutes. When he was 10 years old, he had his first seizure, possibly caused by the accident. As he aged, his seizures grew worse, and eventually medication could not control the attacks. Because the seizures were so severe and incapacitating, Molaison decided to undergo brain surgery.

During surgery, Dr. William Beecher Scoville removed several structures from both sides of Molaison's brain, including the hippocampus, amygdala, and tip of the temporal lobe. Although the surgery significantly reduced his seizures, Molaison was left with a profound memory problem. He was still intelligent and could carry on a conversation, but as soon as he moved on to something new, everything that had just occurred was forgotten. He could remember something if he repeated it over and over in his mind, but as soon as he was distracted, he would forget what he was thinking about. Molaison still had many of the memories of his life prior to surgery, but he could not store any new information or make any new memories. Interestingly, he could learn new motor skills, but he had no memory of how he learned them, where he learned them, or who taught them to him.

When Molaison died on December 2, 2008, his brain

was removed and cut into 2,401 slices to create a three-dimensional digital brain map. This examination confirmed the extent to which the surgery performed back in 1953 altered his brain.

Molaison is sometimes referred to as the most famous subject in neuroscience because the extensive testing of his memory and careful mapping of his brain altered prevailing views about memory. First, it became apparent that there were at least two types of memory: one for facts, names, and dates and another for unconscious acts like riding a bike or tracing a picture. Second, prior to these studies, all parts of the cerebral cortex were thought to contribute equally to memory. The behavioral and neuroanatomical data obtained by studying Molaison showed that the hippocampus and amygdala are critical for the formation of new memories and that memories for skills and facts are stored in different places in the brain.

Neurosurgeons learned an important lesson, too: if you must remove the hippocampus, amygdala, and tip of the temporal lobe, do it on only one side of the brain. It is unfortunate for Molaison that this learning was at his expense.

References

Corkin, S., Lasting consequences of bilateral medial temporal lobectomy: clinical course and experimental findings in H.M., *Seminars in Neurology*, 4:249–59, 1984.

Corkin, S., *Permanent Present Tense the Unforgettable Life of the Amnesic Patient, H.M.* (New York: Basic Books, 2013).

Corkin, S., Amaral, D.G., González, R.G., Johnson, K.A., and Hyman, B.T., H.M.'s medial temporal lobe lesion: findings from magnetic resonance imaging, *Journal of Neuroscience*, 17:3964–79, 1997.

? **?** **?** **?**
Q

Who was Tan?

A

In 1861, French neurologist Paul Broca (1824–1880) examined a 51-year-old man named Louis Victor Leborgne. Leborgne had epilepsy, was partially paralyzed on the right side of his body, and could not speak. Leborgne was given the name "Tan" because that was the only sound he could make. Although Leborgne could understand speech, he had lost the ability to produce it himself.

Soon after an initial examination, Leborgne died. When Broca examined Leborgne's brain, he found damage located on the left frontal lobe. This area is now known as "Broca's area." In 1874, German physician Karl Wernicke (1848–1905) described patients who had left-brain injuries in an area different from that mentioned by Broca. Wernicke's patients had a different speech problem: they could speak, but the words they spoke made no sense. The speech problems associated with brain damage convinced the scientific community that the left side of the brain was critical for language in most people.

The importance of the left hemisphere for language has been confirmed with a procedure called the Wada test. During a Wada test, a barbiturate drug, such as amobarbital (sodium amytal), is injected into the right or left carotid artery. The barbiturate is carried through the bloodstream to the brain on the side the drug was injected. Therefore, the right or left cerebral hemisphere can be shut down independently.

When the left side of the brain is "asleep," the right side of the body becomes paralyzed. People cannot feel anything touching their skin on the right side. The opposite effect occurs when the left side of the brain is asleep: the right side of the body is paralyzed and insensate. In agreement with what Broca saw in Tan, the vast majority of people (96 percent of right-handers, 70 percent of left-handers) who have their left side of the brain anesthetized lose their ability to speak.

The observations that one side of the brain is more important for a particular behavior has given rise to the concept of brain dominance. Even though one side of the brain may be dominant for a particular function, because the right and left sides of the brain are in constant communication with one another, both sides are necessary for efficient, complex, cognitive behaviors.

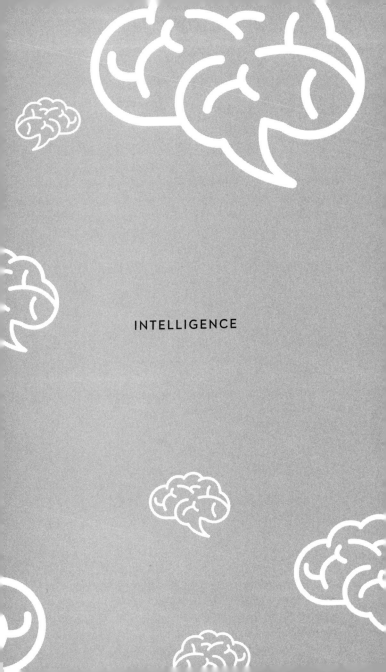

INTELLIGENCE

? **?** **?** **?** **?**

Are there foods that can make people smarter?

A

Your mother was right: a balanced diet is good for you. Eating a variety of foods in moderation benefits your body, including your brain. Food provides the brain with essential vitamins, minerals, and fuel that it needs to operate in top condition. Malnutrition can devastate normal brain development and behavior, but the evidence that particular foods can make you smarter is weak.

The brain uses digested proteins and fats to make synaptic connections and insulate axons with myelin sheaths. Some foods provide the materials to make neurotransmitters. For example, choline (found in eggs) is used to make acetylcholine, glutamic acid (found in flour and potatoes) is used to make glutamate, phenylalamine (found in meat, eggs, and grains) is used to make dopamine, and tryptophan (found in meat, eggs, milk) is used to make serotonin. Fats, particularly n-6 and n-3 fatty acids, are essential for proper brain function because they affect how neurotransmitters are released and how neurons use glucose.

Berries and fish are often touted as "brain foods" that can boost brain power. Blueberries, strawberries, and other fruits and vegetables do contain significant amounts of flavonoids. Flavonoids have antioxidant properties the body uses to eliminate free radicals and reduce inflammation. Free radicals can destroy cells and may lead to neurodegenerative disorders such as Alzheimer's disease and Parkinson's disease. A greater consumption of berries is correlated with fewer signs of cognitive problems later in life. Although this beneficial effect may be related to the ingredients in berries, it also may be related to other factors in people who eat berries. For example, berry eaters may be wealthier, have more time to exercise, or engage in other behaviors that affect brain health.

Fish as brain food has received a great amount of press because it contains significant amounts of omega-3 fatty acids. Omega-3 fatty acids must be obtained through food we eat and can reduce the risk of heart disease. These nutrients are also essential for proper brain development and growth. In fact, babies who do not get enough omega-3 fatty acids in the womb may suffer visual and neurological problems. These results do not imply that a diet high in omega-3 fatty acids will boost your intelligence.

Fish do provide some brainy benefits. People who eat more baked or broiled fish have larger volumes of brain areas (e.g., hippocampus) associated with learning and memory later in life. However, blood levels of omega-3 fatty acids do not correlate with these brain changes. Therefore, eating fish, like eating berries, may be part of a lifestyle that benefits brain health as you get older.

References

Devore, E.E., Kang, J.H., Breteler, M.M., and Grodstein F., Dietary intakes of berries and flavonoids in relation to cognitive decline, *Annals of Neurolology* 72:135–43, 2012.

Raji, C. A., Erickson, K. I., Lopez, O. L., Kuller, L. H., Gach, H. M., Thompson, P. M., Riverol, M., and Becker, J. T., Regular fish consumption and age-related brain gray matter loss, *American Journal of Preventive Medicine*, 47:444–51, 2014.

? ? ? ? ?
Q

Are there drugs that can make people smarter?

A

Actor Bradley Cooper did it on purpose playing Eddie Morra in the movie *Limitless* and actress Scarlett Johansson did it accidentally playing the title character in the movie *Lucy*. What did these characters do? They were both exposed to drugs that gave them incredible memories, astonishing learning abilities, and fantastic cognitive skills. Unbelievable? Well, yes. Remember, it is Hollywood science fiction, not science fact.

Although writers of these and other movies have stretched the science to implausible levels, neuroscientists are interested in how chemicals could be used to enhance cognition. These cognitive enhancers, called "smart drugs" or "nootropics," aim to improve memory, attention, and problem solving. Some drugs attempt to increase brain metabolism or improve cerebral circulation. Other smart drugs try to reduce damage to the brain or alter levels of neurotransmitters. Medication to slow memory loss in people with Alzheimer's disease has shown some promise, but experimental results with drugs used to improve the cognitive abilities of healthy people are weak.

Users of various smart drugs often report that they have improved memory, feel more alert, and can concentrate better. Unfortunately, what people say they feel does not always translate into measurable changes in ability. Furthermore, people who spend money on these drugs are highly motivated to report some benefits. This is why double-blind, randomized controlled experiments are so important: to separate real benefits from perceived improvements. These types of experiments are also needed to uncover undesirable side effects. The long-term consequences of cognitive enhancers on neurotransmitters, neural circuitry, and behavior are not known.

The development of a safe, effective smart drug raises some intriguing philosophical, societal, and legal questions. Who would have access to these drugs, especially if they were expensive? Would using smart drugs before a test be considered cheating? Would we benefit from a society where everyone was "smarter," or would we lose something by creating a population that was more uniform?

Does listening to music make you smarter?

A

Wouldn't it be great if you could boost your intelligence by listening to music? Just play your favorite song and get smarter. Unfortunately, increasing brain power is not so simple.

In the 1990s, researchers performed an experiment that had college students listen to classical music for 10 minutes to see if it could improve their intelligence. The scientists compared the students' spatial reasoning ability after listening to Mozart's *Sonata for Two Pianos in D Major*, after listening to a relaxation tape, or after the students sat in silence. Student test scores improved after listening to Mozart compared to scores after listening to the relaxation tape or sitting in silence. However, this cognitive boost disappeared after 10-15 minutes. This initial experiment prompted other labs to repeat the experiment, but most investigators failed to get similar results. Others studied the possible effects of listening to music on other cognitive measures and also failed to find any beneficial effects. For example, listening to Mozart did not produce better scores on a test for recall of a list of random numbers.

Despite the ambiguous and sparse research results, the perceived ability of music to increase intelligence has come to be known as "the Mozart effect" and spawned a small industry to sell products that probably have no benefit beyond listening pleasure. Even politicians have been hooked by the craze. In the late 1990s, Zell Miller, governor of Georgia at the time, distributed free CDs with classical music to the parents of newborn babies in the state. Perhaps Governor Miller hoped to raise the intelligence of future voters in his state. Unfortunately, there is no evidence to suggest that music influences intelligence permanently, and there have never been any experiments to examine the effects of listening to music on the intelligence of infants.

Listening to classical music may help produce an arts aficionado, but it is unlikely to create a genius.

References

McCutcheon, L.E., Another failure to generalize the Mozart effect, *Psychological Reports*, 87:325–30, 2000.

Ramos, J., and Corsi-Cabrera, M., Does brain electrical activity react to music?, *International Journal of Neuroscience*, 47:351–57, 1989.

Rauscher, F.H., Shaw, G.L., and Ky, K.N., Music and spatial task performance, *Nature*, 365:611, 1993.

Rauscher, F. H., Shaw, G.L., and Ky, K.N., Listening to Mozart enhances spatial-temporal reasoning: towards a neurophysiological basis, *Neuroscience Letters*, 185:44–47, 1995.

Steele, K. M., Brown, J.D., and Stoecker, J.A., Failure to confirm the Rauscher and Shaw description of recovery of the Mozart effect, *Perceptual and Motor Skills*, 88:843–48, 1999.

? ? ? ? ?

Does juggling make you smarter?

A

Learning to juggle can make you the life of the party and it will change your brain, but it is not likely to boost your score on an intelligence test. Just one week of juggling training can expand the visual cortex, and six weeks of juggling can increase the amount of white matter in the brain. If you stick with juggling for three months, the brain areas important for processing information about moving objects can also increase in size.

There probably is not anything special about learning to juggle specifically; people who learn other motor skills such as playing golf or riding a bike have changes in brain structure, too. Regardless of when and where the brain changes after learning a skill, the causes of these changes are not known. More nerve cells, additional glial support cells, or increases in the number of synapses could all increase brain size. There is no evidence that learning to juggle translates to other skills or cognitive abilities.

References

Draganski, B., Gaser, C., Busch, V., Schuierer, G., Bogdahn, U., and May, A., Neuroplasticity: changes in grey matter induced by training, *Nature*, 427:311–12, 2004.

Driemeyer, J., Boyke, J., Gaser, C., Büchel, C., and May, A., Changes in gray matter induced by learning—revisited, *PLoS One*, 3:e2669, 2008.

Scholz, J., Klein, M.C., Behrens, T.E.J., and Johansen-Berg, H., Training induces changes in white matter architecture, *Nature Neuroscience*, 212:1370–71, 2009.

? **?** **?** **?** **?**

What was so special about Albert Einstein's brain?

A

Most people would agree that Albert Einstein (1879–1955) was a smart guy. After all, he came up with the general theory of relativity to explain just about everything we know about light, gravity, and time. He also won the Nobel Prize for Physics in 1921 for his work in theoretical physics.

Since Einstein's death in 1955 from an abdominal aneurysm, people have wondered what made him tick and what was so special about his brain. Soon after Einstein's death, pathologist Dr. Thomas Harvey removed the brain from Einstein's head. Einstein's family did not authorize Dr. Harvey to remove the brain and Einstein had written in his will that he wanted to be cremated. Nevertheless, Harvey took the brain, cut it into 240 small blocks, and kept it in his personal possession for many years. As Dr. Harvey moved from job to job, Einstein's brain moved, too.

Finally, in the 1980s, Dr. Harvey made the famous brain available for scientific study. Einstein's brain was not larger than the average brain. In fact, it weighed only 1,230 grams compared to 1,400 grams, the weight of an average adult male brain. Closer examination of the brain by neuroscientist Marian Diamond and her colleagues revealed that one area of Einstein's brain had a higher ratio of glial cells to neurons compared with other brains. Other researchers found that parts of the cerebral cortex were thinner, and the density of neurons was greater than in comparison brains. The outside of Einstein's brain had an unusual pattern of grooves (sulci) on the parietal lobes. Photographs of Einstein's brain appear to show a larger prefrontal cortex, larger primary somatosensory cortex, larger motor cortex, and thicker corpus callosum than seen on control brains.

These data have been used as anatomical evidence that Einstein's brain was special and somehow bestowed him with superior cognitive abilities. The problem with these studies is that they are all correlative in nature and have only one experimental subject: Einstein. Just because his brain showed some different anatomical characteristics does not mean that these characteristics caused particular behaviors. It is unknown if other people with his level of cognitive prowess have similar external and internal brain structures. Finally, there are no functional data about how Einstein's brain worked—no EEG records or brain imaging data. Examining the size and shape of a brain or the density of neurons and glia provides few clues about how a brain functions.

References

Abraham, C., *Possessing Genius: The Bizarre Odyssey of Einstein's Brain* (New York: St. Martin's Press, 2002).

Anderson, B., and Harvey T., Alterations in cortical thickness and neuronal density in the frontal cortex of Albert Einstein, *Neuroscience Letters*, 210:161–64, 1996.

Colombo, J. A., Reisin, H.D., Miguel-Hidalgo, J.J., and Rajkowska G., Cerebral cortex astroglia and the brain of a genius: a propos of A. Einstein's, *Brain Research Reviews*, 52:257–63, 2006.

Diamond, M. C., Scheibel, A.B., Murphy, G.M. Jr., and Harvey, T., On the brain of a scientist: Albert Einstein, *Experimental Neurology*, 88:198–204, 1985.

Falk, D., Lepore, F.E., and Noe, A. The cerebral cortex of Albert Einstein: a description and preliminary analysis of unpublished photographs, *Brain*, 136:1304–27, 2013.

Paterniti, M., *Driving Mr. Albert: A Trip Across America with Einstein's Brain* (New York: Dial Press, 2000).

Witelson, S. F., Kigar, D.L., and Harvey, T., The exceptional brain of Albert Einstein, *Lancet*, 353:2149–53, 1999.

Do you get a new wrinkle in the brain every time you learn something new?

A

What exactly is a "wrinkle?" The surface of the brain does have bumps (gyri) and grooves (sulci). These folds increase the amount of surface area that can fit in the volume of the skull. Although all brains have the same general pattern of folding, each one is unique with its hills and valleys taking different shapes and sizes.

Learning is thought to occur when the connections between neurons become stronger. This strengthening of the synaptic connections between neurons makes it easier for them to communicate with each other. This is a physical change in the brain that takes place each time learning occurs. If you want to call synaptic changes "wrinkles," then yes, they do happen every time you learn. But, no, learning does not result in new bumps or grooves on the surface of the brain.

Does watching TV, playing video games, or surfing the Internet kill brain cells?

A

Soap operas, reality TV, the latest video game, and amazing websites won't do much to boost your intelligence, but they also won't kill brain cells directly. All electronic devices emit electromagnetic radiation, but the amount of this energy directed into your brain from a screen is relatively small, especially if you are watching a show or playing a game from a few feet away from the screen.

Sitting in front of a screen is not good for you. If you have a box of cookies, a can of soda, and a bag of potato chips as you watch or play, you are not doing your body any good. Reading a book or getting some exercise is certainly better than staring at a screen.

References

Austrew, A., The AAP finally admits screen time isn't the enemy, accessed January 30, 2016, http://www.scarymommy.com/the-aap-finally-admits-screen-time-isnt-the-enemy/.

Brown, A., Shifrin, D.L., and Hill, D.L. Beyond "turn it off": How to advise families on media use, AAP News, September 28, 2015, http://www.aappublications.org/content/36/10/54.full.

MEMORY

Is memory like a tape recorder, flash drive, or hard drive?

<u>A</u>

In a word, no. Even if you think you have a great memory, your memory for times, places, and events is not very good. There are several reasons for this. First, what we observe is not the same as what exists. Part of the issue is that our sensors are not perfect. We have a blind spot in our vision, for example. However, there is another, more subtle reason for this discrepancy, and that is that we have a lot of cognitive machinery in between the real world and our perception. Things like attention, expectation, and emotional arousal will change what we "see." Our brains use autocomplete all the time, which is why you don't see (or not see) your blind spot. Even before your experiences get to your memory, they have diverged from objective reality. But there is another issue that has to do with the way memories are stored.

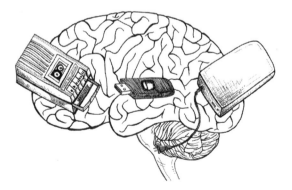

When your brain initially stores a memory it is called *consolidation*: the process of transferring an experience into long-term storage. Basically, the pieces (sensations, emotions, etc.) of a memory are bound together so that later you can revisit that experience. When this is accomplished, the memory is stable. But this isn't the end of the story. Whenever you recall that memory, you bring it back into a labile (changeable) state, and you have to consolidate it again. In the process of that reconsolidation, you include a little bit about the state you were in when you recalled it. This means that every time you recall a memory, you change it just a little bit. Perversely, the memories we recall the most frequently, probably our most important memories, are the least accurate.

This seems like a bummer, and in some ways it is. However, this isn't a case of our brains not being very good. You have to consider what your memory has evolved for, and it turns out that it isn't to provide you with a detailed and accurate record of your experiences. Our memories help us use previous experience to optimize future actions, and for this, the gist of an experience is usually sufficient. There is usually some reason you recall a certain memory, and that is important information, too. Reconsolidation allows you to store this new information with the original memory, which is efficient.

The real problem is that we feel like our memories are very accurate. Unfortunately, this feeling is very misleading. The vividness of a memory and the confidence we have that it is true do not correlate with accuracy. This can have serious consequences in the context of, say, eyewitness testimony.

Our memories give us a sense of where we have come from and who we are. In this sense, memories true. But sometimes that is the only sense in which they are true.

Can memories be erased?

A

Yes. Memories are represented and stored in the brain as a set of connections between neurons. These connections, because they are cellular in nature, rely on proteins. If you can't make these proteins during the period of memory formation (or consolidation), then you can't create a memory. Chemical agents called protein inhibitors can prevent protein synthesis. By delivering these chemicals, scientists have been able to prevent the creation of new memories. This is not the same as erasing existing memories, but its close.

Now we get to a very important point: *reconsolidation*. Although many people believe that memories are permanent and unchanging, they are not. Every time a memory is recalled, it returns to an active, malleable state and must be rewritten into the brain. As scientists say, it has to be *reconsolidated*. This means that every time a memory is recalled, there is a window when, just like a new memory, it can be tampered with. If given at the right time, the very same protein inhibitors that prevent consolidation of a new memory will prevent reconsolidation of an existing memory. In other words, the memory will be destroyed. Whether this is a good thing or a bad thing depends on your perspective.

Reference

Nader, K., Schafe, G.E., and Le Doux, J.E., Fear memories require protein synthesis in the amygdala for reconsolidation after retrieval, *Nature*, 406:722–26, 2000.

? **?** **?** **?** **?**

Can memories be implanted?

A

Yes, they can, and it isn't even all that difficult. No pharmaceutical, electrical, surgical, or other invasive measures are required. It is possible that someone has done it to you without either of you knowing it. We know this largely due to the pioneering work of psychologist Elizabeth Loftus (1944–).

Loftus started investigating the idea in the early 1990s when many people were recovering memories of extreme and bizarre childhood abuse through psychotherapy. To her, something seemed not quite right. She already knew that memory was vulnerable to suggestion and began to wonder if it was possible to suggest a memory into existence. In a now famous experiment, Loftus and her students asked 24 subjects to recall childhood events, the details of which were supplied by close relatives. Unbeknownst to the subjects, three of the events were true and one was false. After reading the memories and then being asked to recall them on two separate occasions, six of the subjects said that they remembered the false incident. When told that one of the memories was false and asked to decide which one it was, five of the subjects chose incorrectly. This was a small study, but the results have been replicated over and over again with slight variations. In some studies, researchers have been able to implant false memories in up to 50 percent of their test subjects.

Loftus and her colleagues have proposed a three-process model for implanting a false memory. First, you have to make the event seem plausible. It turns out that even very unlikely things can be made to seem plausible if someone you trust tells you so. Next, you make it seem plausible that the event happened to the subject. Finally, you help the subject interpret thoughts and fantasies about the subject as memories. This was exactly what was happening in repressed memory therapy. This does not mean that no one has ever repressed a memory of childhood abuse. It means that therapists need to proceed with extreme caution and avoid the suggestive techniques outlined here.

You may have never participated in repressed memory therapy, but you may still have false memories. It is possible, and even probable, that many of your childhood memories were implanted by your parents. They aren't false memories in the sense that they didn't happen. Rather, they are false in the sense that you don't really remember them. But if you hear a story and see the photos enough times, you might think that you do.

How can you tell the difference between real memories and fake memories? Without an objective reference, you can't. If objective history is important to you, it's not a bad idea to keep a diary.

Reference

Mazzoni, G.A.L., Loftus, E.F., and Kirsch, I., Changing beliefs about implausible autobiographical events: a little plausibility goes a long way, *Journal of Experimental Psychology: Applied*, 7:51–59, 2001.

Why can't I remember being a baby?

A

Our first experiences must be some of the most salient, precisely because they are the first. What does it feel like to eat solid food for the first time, to take your first steps, to have your first cold, to say your first word, or to learn to use the toilet? We used to know, but we have forgotten. This phenomenon is called infant amnesia or childhood amnesia. Humans aren't the only species that experience this phenomenon; other mammals that are born immature also forget their very early memories.

The problem is not that young children do not form memories, but that they don't hang on to them. Most people remember very few events from before they were 3 years old, remember a few more events for every year between 3 and 7, and between 7 and 10 reach a plateau. Whatever else it may be, this phenomenon is clearly related to development. Several theories have been advanced to explain this forgetting. One theory has recently been bolstered by substantial experimental evidence. According to this theory, the culprit is neurogenesis (the creation of new neurons) in the hippocampus. The hippocampus is a part of the brain that is particularly important for forming episodic memories, and neurogenesis continues here throughout life. This is a good thing because it allows us to form new memories. Unfortunately, these new neurons compete with the older ones, sometimes shoving older memories out.

Hippocampal neurogenesis is at its highest during infancy and then tapers off. This could explain why we have a graded effect for childhood memories and why our adult memories are more resistant. This is a shame because many of us would like to remember the sense of wonder and joy that children seem to experience. Of course, they likely also experience a lot of confusion and frustration. In any case, the parents of very young children are often too sleep-deprived to remember things clearly. So by the time we get to adulthood, most of our early childhood has been lost.

References

Akers, K. G., Martinez-Canabal, A., Restivo, L., Yiu, A.P., De Cristofaro, A., Hsiang, H.L., Wheeler, A.L., Guskjolen, A., Niibori, Y., Shoji, H., Ohira, K., Richards, B.A., Miyakawa, .T, Josselyn, S.A., and Frankland, P.W., Hippocampal neurogenesis regulates forgetting during adulthood and infancy, *Science*, 344:598–602, 2014.

Josselyn, S. A., and Frankland, P.W., Infantile amnesia: a neurogenic hypothesis, *Learning & Memory*, 19:423–33, 2012.

? ? ? ? ? ## Q

Does a bump on the head cause a person to lose memories, and does another bump on the head cause the person to remember things?

A

You may have seen a movie or a cartoon where a character gets into an accident or hit on the head and then can't remember what happened. Sometimes these characters suffer a second head injury and magically remember everything again. Traumatic brain injury (TBI) is a significant health risk worldwide, and sometimes it does affect memory. But it is a myth that a second blow to the head will restore memory.

In the United States alone, 1.7 million people suffer from a TBI, and 52,000 people die of these injuries each year. The symptoms of a TBI range from mild to severe depending on the location and extent of the injury. Symptoms can include changes in thinking, remembering, mood, and sleep. Many people who have suffered a TBI complain of headaches, blurred vision, nausea, dizziness, and sensitivity to light. The loss of consciousness is not required for the diagnosis of a TBI. Rather, any change in mental status could indicate an injury to the brain. Some head injury symptoms last only a few seconds or minutes, but others can last days, months, or even years.

Despite what filmmakers would have you believe, it is extremely unlikely that another impact to the head will restore these memories. Another knock to the noggin will more likely result in another, more severe brain injury. Head injuries can affect memory: people who have a TBI report that they cannot remember specific events in their past or that they have trouble forming and storing new memories. Memory problems, dizziness, headaches, insomnia, depression, attention problems, and other neurological disorders are a few of the symptoms experienced by football players who suffered repeated concussions. Football players who have had multiple mild TBIs show evidence of chronic traumatic encephalopathy, a disorder characterized by degeneration of the frontal and temporal lobes of the brain and other neurological abnormalities. These findings have motivated the National Football League to fund research into new technologies to prevent head injuries and add rules to limit the incidence of head-to-head contact.

References

Faul, M., Xu, L., Wald, M.M. and Coronado, V.G., Traumatic brain injury in the United States: emergency department visits, hospitalizations, and deaths, Centers for Disease Control and Prevention, National Center for Injury Prevention and Control, Atlanta, GA, 2010.

McKee, A.C., Stein, T.D., Nowinski, C.J.,Stern, R.A., Daneshvar, D.H., Alvarez, V.E., Lee, H-S., Hall, G., Wojtowicz, S.M., Baugh, C.M., Riley, D.O., Kubilus, C.A., Cormier, K.A., Jacobs, M.A., Martin, B.R., Abraham, C.R., Ikezu, T., Reichard, R.R., Wolozin, B.L., Budson, A.E., Goldstein, L.E. Kowall, N.W., and Cantu, R.C., The spectrum of disease in chronic traumatic encephalopathy, *Brain*, 136:43–64, 2013.

SLEEP

Why do we sleep?

A

William Dement, a pioneer in the field of sleep research, famously said, "the only reason we need to sleep that is really, really solid is because we get sleepy." We can go one better than Dement and say that we really need to sleep because if we didn't, we would die. If you don't find either of these answers very satisfying, get in line.

Sleep is one of the most interesting and mysterious things that humans do. We spend about one third of our adult lives asleep, in a state of altered consciousness, vulnerable to predation. When we wake up, we remember very little of what happened. What we do remember is deeply strange. On the surface, this doesn't seem to be a good use of our limited life span. But we have to do it. In fact, all animals (even insects) have to do it. The need for it is so great that if you are sufficiently sleep-deprived, the drive to rest will become greater than the need for food and water. Sleep deprivation can be a form of torture. You can be sure, then, that something very important is happening during sleep. Probably many very important somethings are happening during sleep.

One of the ways we can determine the function of sleep is to see what happens to people when they don't get it. It turns out that a lot of things go downhill very quickly. Basically, if you don't sleep enough you will be sick, overweight, stupid, emotionally unstable, and a public hazard. Most people are familiar with this, at least on some level. As noted, if you don't sleep at all, you will die. This is exactly what happens to people with a very rare disease called fatal familial insomnia. It isn't a nice way to die (if there is a nice way to die).

It is clear that sleep is particularly important for brain function and immune function, but scientists are still working out all of the ways that is true. There are different stages of sleep, and those different stages likely have different physiological functions. For example, we know that some stages of sleep are very important for learning, and others appear to be more important for creativity. The take-home message is that sleep is something you need to do.

? ? ? ? ?

Why do we dream?

A

Dreams are weird. They don't make a lot of sense, can be highly emotionally charged, are often anachronistic, and are usually hard to remember. On the other hand, some dreams are so striking that we remember them for years afterward. Some dreams play over and over, night after night. Intuitively, dreams feel like they should be important—like they convey some information to which we don't have access while we are awake.

Since ancient times, dreams have been treated as a form of special revelation, sometimes from a divine source and sometimes as a message from our subconscious. In both cases some interpretation is usually required—and there is a lot of disagreement about the right way to do this (although sometimes a cigar is just a cigar).

What can science tell us about the purpose of our dreams? Not much. First, there are different types of dreams: those that happen during rapid eye movement (REM) sleep, and those that happen during non-REM (NREM) sleep. The REM dreams are the really vivid, bizarre ones. NREM dreams tend to be a bit repetitive and boring. This is consistent with one of the substantiated functions of NREM: to replay the day's events so that they can be transferred into long-term memory. REM dreams are trickier because no one really knows what REM is for. This is interesting, but also frustrating.

Many theories attempt to explain REM sleep (some more scientific than others), but there are few solid answers. Some of the more legitimate options include memory formation and consolidation, creative problem solving, and behavioral rehearsal (these are not necessarily exclusive functions). Some sleep researchers have ventured to say that REM dreams have no function, although this seems premature—just because we haven't discovered the function of something doesn't mean that no function exists. Dreaming may be important. But if there is an adaptive function of dreaming, we don't know what it is. So if someone tells you they can interpret your dreams, know that they are not using an evidence-based process and proceed at your own risk.

? ? ? ? ?
Q

How much sleep do I need?

A

It depends on how old you are. Infants need far more sleep than their parents do (thank goodness). But assuming you are an adult (which you are if you're 18 or older, even if you still do laundry at your parents' house) then you need about seven hours a night. If you sleep much more or much less than that, then your risk of all-cause mortality (your chance of dying of something, but not any one thing in particular) is significantly increased. Those in the "I'll sleep when I'm dead" camp will likely end up sleeping longer than they want. That's not so surprising, because we all know that when we don't sleep we feel bad. Feeling bad is nature's way of telling you that whatever you just did was the wrong thing to do.

What is more surprising to many people is that sleeping too much can be equally bad for you. In that case, most of your abbreviated life will be spent unconscious. It should be noted, however, that causality has not been shown here. This means that we don't know if people are sleeping more because they're dying, if they're dying because they're sleeping more, if dying and sleeping are both caused by an unknown third factor, or if the correlation is just incidental. With all of that said, try to sleep about seven hours a day.

You should also try to get your seven hours of sleep at night, if at all possible, because that's when your body wants to be asleep. Thanks to your circadian clock, your body knows what time of day it is (unless you never see the sun), and it wants to be awake during daylight hours. Humans are diurnal: you will be healthier if you accept and live at peace with this fact.

There are some people who are exceptions to the seven-hour rule. These super-sleepers seem to need only four or five hours of sleep to function well. This is likely due to a genetic mutation. If you aren't a super-sleeper, you won't become one by training. If you are one of the lucky few who wakes up cheerful and alert after only a few hours of rest, count your blessings and do something productive and make a pot of coffee for the rest of us.

References

National Heart, Lung, and Blood Institute, How much sleep is enough?, February 12, 2012, accessed February 1, 2016, http://www.nhlbi.nih.gov/health/health-topics/topics/sdd/howmuch.

National Sleep Foundation, How much sleep do we really need?, accessed February 1, 2016, https://sleepfoundation.org/how-sleep-works/how-much-sleep-do-we-really-need.

? **?** **?** **?**
? **?**

Will studying all night improve my test scores?

A

As with many things in life, the answer is, "it depends." It depends on the kind of test, how much studying you have already done, and whether you would like to actually retain the information or just regurgitate it and forget it.

Most people have, at some point in their lives, found themselves less prepared than they want to be for a test or exam. When this happens there is often a temptation to spend every remaining minute, even minutes when you would usually be sleeping, trying to cram information into your head. This is typically not a great strategy. For one thing, acute sleep deprivation impairs your ability to pay attention, recall facts, make decisions, be creative, and reason through problems. If you're taking an exam that will require you to think in any meaningful way, you're not doing yourself any favors by pulling an all-nighter. It is also true that your studying will become less and less efficient as you become more tired, so you're not likely to learn all that much between three and four o'clock in the morning anyway. In addition, you need to sleep to consolidate learning, so if you cram your head full of facts without sleeping on them, you're unlikely to remember them later.

Even so, sleeping will not allow you to magically learn things you never studied. If you are starting with a knowledge base of zero, then you will probably benefit (in terms of your exam score) by sleep depriving yourself a little and hitting the books. If you have already studied some, then you'll probably do better if you go to bed. Theoretically there is a sweet spot where the trade-off between studying and sleeping tips, but since everyone is different and every test is different, you'll have to estimate this point for yourself. If you find you are staring at the wall more than you are studying, or if you are falling asleep in your book, you have passed the sweet spot and should go to bed. If you have already studied for many hours and simply do not understand the material, then you should definitely go to bed, especially if you are starting to feel emotionally agitated.

Of course, if you really want to do well on exams you should start studying several days in advance and give yourself the chance to sleep on it several times.

? ? ? ? ?

Q

Can you learn while you're asleep?

A

Yes, but probably not in the way you think (or hope) you might. To many of us, sleep seems to be a time when we're not doing much, so it is attractive to think we could put all that wasted time to good use by learning something. This is especially appealing if you find whatever you're hoping to learn boring or otherwise unpleasant—so much the better if you can be unconscious while you learn it! Consequently, students have been sleeping on books or playing audio recordings to their sleeping selves for many years in hopes that they will wake up speaking Spanish or have memorized 100 lines of Shakespeare. Unfortunately, this doesn't really work (and so a beautiful dream dies).

Although it looks like you aren't doing anything while you're sleeping, your brain is actually quite busy. Ironically,

one of the things that your brain is doing is consolidating—firming up—all of the learning you did while you were awake. During sleep your brain is busily moving things from short-term memory to long-term memory, and part of what your brain needs to do this is a little peace and quiet. That is, your brain needs to be protected from new experiences while it is working on storing the old ones. Your brain shuts the gates to incoming sensory information, unless that information is especially strong (like a bucket of cold water) or especially meaningful (like the sound of your name being spoken). Sensory inputs of that nature will terminate your sleep. If you can't get sensory information into your brain, you can't really learn anything new.

However, there are a few loopholes. First, it is only *mostly* true that sensory information doesn't get into your cerebral cortex. Some sounds that are below threshold for waking will be passed through by the thalamus, and smells can always get into the cortex because olfaction is the only sense that bypasses the thalamus entirely (fun fact). Several researchers have taken advantage of this fact to boost consolidation of particular memories. The idea is that during wakefulness certain sounds or smells are associated with a particular task. During postlearning sleep, these sounds or smells are replayed and the associated task is preferentially consolidated, meaning it is learned better than an equivalent task that wasn't cued during sleep. So you can basically tell the brain what you want it to remember, which is neat.

In addition, researchers recently validated years of seemingly senseless hoping when they showed that in fact, humans can unconsciously learn something new while they sleep. To do this, they used the fact that when people smell a pleasant odor they sniff more strongly than when they smell an unpleasant odor. They paired unpleasant and pleasant odors with different tones and then played the tones without the odors to see what kind of sniffs they got. If the subjects sniffed in response to the tones alone, then learning would have occurred (think of Pavlov's dogs). They sniffed in their sleep, and they still sniffed when they were awake, without any awareness that they were doing it. This is amazing, but also rather limited. So if you want to learn a new language, you're still going to have to do it the old-fashioned way.

Reference

Antony, J.W., Gobel, E.W., O'Hare, J.K., Reber, P.J., and Paller, K.A., Cued memory reactivation during sleep influences skill learning, *Nature Neuroscience* 15:1114–16, 2012.

What is sleepwalking?

There are two phenomena that can be called sleepwalking—somnambulism and rapid eye movement (REM) sleep behavior disorder. Somnambulism occurs early in the night in the deep stages of non-REM sleep and is usually not dangerous. It is sometimes even funny. The sleeper sits up or rises from bed, sometimes with eyes open, sometimes talks, sometimes walks around, or goes to another room. The episodes usually last about 10 minutes, and the sleepwalker does not remember it on waking. If you wake the sleeper during an episode, he may become disoriented or even aggressive. Waking up the sleepwalker will not harm him. In fact, it is sometimes imperative to wake a sleepwalker. Sleepwalkers may get lost, may lose their balance and fall, and may return to normal sleep in strange places or positions. Some people have been known to cook or drive in their sleep, which presents an obvious danger.

Sleepwalking runs in families, and most sleepwalkers are children. Sleepwalking can be a symptom of other neurological or psychological problems, but it usually isn't. If a child occasionally gets up in his sleep and uses the closet as a toilet, it's annoying, but not necessarily something to worry about.

REM sleep behavior disorder (RBD), on the other hand, is a different animal. In most people, movement is suppressed during REM sleep, the time when you experience vivid and bizarre dreams. If you have RBD, you are not immobilized during REM, and you act out your dreams. This is not good because dreams are often terrifying or emotionally charged, and some have a chasing or attacking theme. In the middle of such an episode, a person with RBD might leap out of bed, run, punch, kick, bite, scream, or curse. People with RBD are often injured when they run into or hit things or when they fall. This can be very bad, not to mention frightening. It's also very dangerous for anyone who might be sleeping nearby who may be hurt or killed when they are mistaken for someone in the dream. RBD may be associated with a neurodegenerative disease or it may be idiopathic, meaning no one knows why it happens. Regardless of why it is happening, if you think it is happening to you, you should visit your doctor.

References

Boeve, B.F., REM sleep behavior disorder: updated review of the core features, the RBD-neurodegenerative disease association, evolving concepts, controversies, and future directions, *Annals of the New York Academy of Sciences*, 1184:15–54, 2010.

Dugdale, D. C. III, MedlinePlus, Sleepwalking, April 14, 2013, accessed January 30, 2016, https://www.nlm.nih.gov/medlineplus/ency/article/000808.htm.

? **?** **?** **?** **?**

Q

What is lucid dreaming?

A

The simple answer is that lucid dreaming is being conscious of the fact that you are dreaming while you are still asleep. Unfortunately, there are no more easy answers on this topic. What we want to know is what is happening in the brain when people are lucid dreaming. This is very difficult to know because people are unresponsive when they're doing it. Are they really asleep, awake, or somewhere in between? This is a matter of some debate. Here are the things we know about lucid dreaming:

- It arises from the sleep state characterized by vivid, bizarre dreams, and rapid eye movement (REM).

- Brain activity is different during lucid dreaming than it is during normal REM sleep.

- People can sometimes signal that they have become lucid in a dream with a prearranged series of eye movements.

- Lucid dreaming can happen spontaneously, but people can train themselves to do it with practice.

- Lucidity can be induced by an experimenter using electrical stimulation.

Because the function of normal REM dreams is unknown, it is difficult to know what the function of lucid dreaming might be, if it has any function. Lucid dreaming can be used to treat people who suffer from nightmares, and some practiced lucid dreamers use it to find inspiration or solve problems.

THE SENSES AND PERCEPTION

How many senses do we have?

A

Most people can name touch, vision, hearing, taste, and smell as our senses. But humans have more than these five senses, and other animals have senses that humans do not have.

Our senses bring in information about the world around us and the world inside our body. Given this definition, we have more than the five senses. Let's start with the sense of touch. The skin has specialized receptors that respond to different types of stimulation such as stretch, vibration, pressure, temperature, and pain. These receptors are connected to different axons that send information about what is happening to the skin. Even vision could be divided into two separate senses—color vision and brightness—because of the different types of receptors in the retina that respond to color (cone receptors) and brightness (rod receptors).

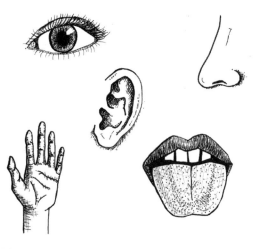

Some people might argue that defining the number of senses by differences in the types of receptors may not "make sense." For example, receptors for taste (sour, sweet, bitter, salty, umami) and smell (thousands of receptors) are tuned to respond best to chemicals with specific shapes. Do we have five different taste senses and thousands of different smell senses?

Regardless of how a sense is defined, there are certainly other forms of sensation that do not fall into the touch, vision, hearing, taste, and smell categories. The sense of balance, provided by receptors in the semicircular canals of the inner ear, is a good example. We also have special cells in our muscles, joints, and tendons that provide information about our bodies in space (proprioception). Even our internal organs (heart, stomach, lungs, kidneys) have receptors that help regulate body processes such as hunger and thirst. These internal senses also tell us when something is wrong, for example, when we feel the pain of a kidney stone or heart attack.

The realm of the senses might even extend to the ability to sense time. Although specific receptors to judge the passing of time have not been identified, time provides us with important information about our environment.

Other animals have the ability to detect some forms of energy that humans cannot. For example, sharks, eels, and rays have special receptors that respond to electrical fields and some turtles and birds use magnetic fields to navigate. Animals can also detect signals beyond the range of humans: bats, dolphins and dogs can hear high frequencies of sound, some butterflies and birds can see ultraviolet light, and pit vipers can detect infrared radiation.

? ? **?** ? ,

Q

Why are my fingers and face more sensitive than my legs and back?

A

Your skin is the largest organ of your body, weighing approximately 9 pounds (4.1 kilograms) and covering 3,000 in² (1.8 m²) of your frame. This thin sheet of tissue keeps your insides in, shields you from injury and germs, helps regulate your body temperature, and provides information about touch, pressure, pain, chemicals, cold, heat, and vibration.

The skin contains millions of sensory detectors (receptors) sensitive to different types of stimuli. For example, some receptors respond to vibration, and others respond to heat. Each sensory receptor is connected to an axon that sends information about the stimulus into the central nervous system, up to the brain. In the brain, information from the skin eventually reaches the primary somatosensory cortex, an area in the parietal lobe of the cerebral cortex. As information reaches this area, it is represented in an orderly manner according to where in the body it originated. In other words, a map of the body is created in the brain. For example, neurons in the cerebral cortex that receive information from the hand will be located near neurons that respond to stimulation of the arm.

The brain map of the body surface is distorted, however. The map is not an accurate representation of the amount of skin devoted to a particular body part. Instead, the amount of cerebral cortex allocated to the skin of a particular body area is proportional to the density of sensory receptors in the skin. The skin covering the fingers, hands, and face has many more sensory receptors than the skin on the back, arms, and legs. Because there are more neurons in the cerebral cortex for processing information from the fingers, hands, and face, we are more sensitive to touch, pressure, pain, and temperatures applied to those areas of our bodies.

You can easily examine the sensitivity of your skin with a simple two-point discrimination test. First, bend a paper clip into a U shape. Then touch different areas of your skin with both ends of the paper clip at the same time to check if you can feel two distinct points or only one. Vary the distance between the paper clip ends to see if the smallest distance where two points are distinctly separate changes for different parts of your body.

? **?** **?** **?** **?**

Would I feel anything if my brain was touched?

A

Push it, pull it, poke it, stretch it: no matter how you touch the brain, a person will not feel it. That's because the brain does not have any sensory receptors for touch. The brain is the organ for the perception of touch from other parts of the body, but it does not have any sensory receptors to provide information about itself.

The linings that cover the brain (the meninges: dura mater, arachnoid, pia mater) do have sensory receptors. The sensory information from the meninges is carried in the trigeminal nerve, vagus nerve, and upper cervical spinal nerves to the brain. Some headaches are thought to be caused by activation of small nerve fibers within the meninges.

Although touching the brain will not cause any feeling, electrical stimulation of the brain will result in a perception. Neurosurgeons take advantage of this phenomenon when they map the brain during surgery, such as when a brain tumor is removed. Surgeons must be careful to remove only diseased tissue, so they apply small amounts of electrical current to the brain to see how a patient responds. Electrical stimulation of brain areas responsible for the perception of touch, for example, the somatosensory cortex, will cause a person to feel a specific part of the body. The sensation is often described as "wind running down the hand," "light rub or a light buzz," "muffled," or "something was wrapped around" the finger. Stimulating areas of the brain responsible for speech will affect the way a person talks. Neurosurgeons are careful not to damage these critical areas as they go about their work.

References

Fricke, B., Andres, K.H., and Von Düring, M., Nerve fibers innervating the cranial and spinal meninges: morphology of nerve fiber terminals and their structural integration, *Microscopy Research and Technique*, 53:96–105, 2001.

Johnson, L.A., Wander, J.D., Sarma, D., Su, D.K., Fetz, E.E., and Ojemann, J.E., Direct electrical stimulation of somatosensory cortex in humans using electrocorticography electrodes: a qualitative and quantitative report, *Journal of Neural Engineering*, 10:036021, 2013.

Do insects feel pain?

A

Do you cause unnecessary pain and suffering to a mosquito when it gets buzzed in the bug zapper? Or is your conscience clear because bugs do not feel pain? According to the International Association for the Study of Pain, pain is "an unpleasant sensory and emotional experience associated with actual or potential tissue damage, or described in terms of such damage." This definition implies that pain is more than just having sensory receptors that respond to tissue damage.

Insects respond to obvious injuries in a very different way than do humans and other animals. For example, an insect with a broken leg or wing still feeds, mates, and behaves normally. Some insects, like cockroaches, can actu-

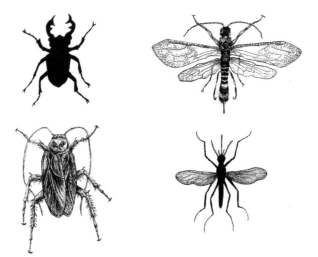

ally live without their heads for days or weeks. Even when their heads are attached, insects don't have the higher brain areas that are needed to interpret the emotional and cognitive significance of a painful stimulus.

So what does a fly or a mosquito feel as you swat it away? We really don't know. But even though insects may react and withdraw from a damaging stimulus, it is unlikely that they experience pain, at least not in the way humans experience pain.

Reference

Classification of Chronic Pain, 2nd ed., IASP Task Force on Taxonomy, edited by H. Merskey and N. Bogduk (Seattle: IASP Press, 1994).

? ? ? ? ?

Do all people experience pain?

A

You might think that life without pain would be great: no more loud cursing when you bump your knee or get stung by a wasp. But pain is important because it protects you from further injury. Without pain, you might not remove your hand from a hot stove or take care of a broken arm or rotten tooth. Without treatment, you might suffer further injury and get a serious infection.

People with a rare condition called congenital insensitivity to pain feel no pain. They can be jabbed with a pin, cut with a knife, or bite their own tongue and go on their merry way without showing any discomfort. The first signs that someone may have congenital insensitivity to pain often occur early in life when a baby does not show any distress after getting an injection, having blood drawn, or injuring an eye. Some children show no response after suffering what should be a painful burn.

Congenital insensitivity to pain appears to be an inherited disorder. People born with this condition have a mutation that alters the transport of sodium ions across channels in sensory nerves. Because these channels do not work, sensory nerves do not provide information about pain to the brain. Therefore, signals about burns, ulcers, and broken bones never reach the brain and a person will not take action to stop the tissue damage. Unfortunately, people with congenital insensitivity to pain have a reduced life span because of the repeated injuries and subsequent health problems they suffer.

References

Chen, Y.-C., Auer-Grumbach, M., Matsukawa, S., Zitzelsberger, M., Themistocleous, A.C., Strom, T.M., Samara, C., Moore, A.W., Cho, L.T-Y., Young, G.T., Weiss, C., Schabhutti, M., Stucka, R., Schmid, A.B., Parman, Y., Graul-Neumann, L., Heinritz, W., Passarge, E., Watson, R.M., Hertz, J.M., Moog, U., Baumgartner, M., Valente, E.M., Pereira, D., and Restrepo, C.M., Transcriptional regulator PRDM12 is essential for human pain perception, *Nature Genetics*, 47:803–8, 2015.

Golshani, A.E., Kamdar, A.A., Spence, S.C., and Beckmann, N.M., Congenital indifference to pain: an illustrated case report and literature review, Journal of Radiology Case Reports, 8:16–23, 2014.

Minett, M.S., Pereira, V., Sikandar, S., Matsuyama, A., Lolignier, S., Kanellopoulos, A.H., Mancini, F., Iannetti, G.D., Bogdanov, Y.D., Santana-Varela, S., Millet, Q., Baskozos, G., MacAllister, R., Cox, J.J., Zhao, J., and Wood, J.N., Endogenous opioids contribute to insensitivity to pain in humans and mice lacking sodium channel Nav1.7, *Nature Communications*, 6, doi:10.1038/ncomms9967, 2015.

Nagasako, E.M., Oaklander, A.L., and Dworkin, R.H., Congenital insensitivity to pain: an update, *Pain*, 101:213–19, 2003.

What is phantom limb pain?

A

After someone loses a limb, for example, following an amputation, the limb is physically gone but its presence often remains in the form of phantom sensations. These sensations include an awareness of the limb, the feeling that it is in a strange position, tingling, itching, and, in a classic case of adding insult to injury, pain. This pain can be extreme, debilitating, persistent, and almost impossible to treat.

One reason that phantom limb pain is so difficult to control is because the source of the pain is not obvious. Certainly there are severed nerve fibers in the stump, but a series of failed treatments have made it clear that this is not the whole story of phantom pain. Instead, it appears that the larger problem is not in the stump but in the brain. Specifically, the brain has an existing representation for the expected sensory inputs from and the motor outputs to the limb. When these expectations are not matched to the visual feedback (which stubbornly insists that the limb is missing), the result is pain. The (at least partial) proof for this theory is in the pudding: experimental new treatments designed to resolve this mismatch are effective at reducing pain.

One of the first methods targeted at the perceptual discrepancy was a mirror box task in which subjects saw their amputated arm as the mirror image of their remaining, healthy arm. It sounds strange, but even though subjects knew that they were looking at a visual illusion, they began to perceive the illusory arm as their own. In this state of reembodiment, subjects were able to move their existing limb and perceive it as movement of their missing limb. In this way people could unclasp their phantom hand or move their phantom arm out of a painful position, thereby reducing their experience of pain. Other reembodiment strategies have also been helpful. However, more recent research has shown that targeted injections of anesthetics in the peripheral nervous system can also dramatically diminish pain. All of this would suggest that the jury is still out on exactly what causes phantom limb pain.

References

Ramachandran, V.S., and Rogers-Ramachandran, D., Synaesthesia in phantom limbs induced with mirrors, *Proceedings of the Royal Society of London B: Biological Sciences,* 263:377–86, 1996.

Vaso, A., Adahan, H.M., Gjika, A., Zahaj, S., Zhurda, T., Vyshka, G., and Devor, M., Peripheral nervous system origin of phantom limb pain, *Pain,* 155:1384–91, 2014.

Wellcome Trust, Phantom limb pain, http://www.wellcome.ac.uk/en/pain/microsite/medicine2.html, accessed January 26, 2016.

Why are teeth so painful?

A

Many people dread a visit to the dentist. But dentists are our friends, especially when we have a toothache.

The anatomy of a tooth is really a work of art. Each tooth consists of several layers. The outermost layer is the enamel, the strongest material in the body. Enamel does not have any nerve cells. Below the enamel is the dentin. The dentin is filled with many small tubules containing fluid and special cells including nerve fibers. Those fibers are branches from sensory axons in the center of the tooth, the tooth pulp. Sensory axons connect to their cell bodies near your brain via the trigeminal nerve.

Small myelinated nerve fibers and small unmyelinated nerve fibers make up the vast majority of axons in the tooth. These two types of axons are responsible for different pain sensations in the teeth. The small myelinated nerve send sharp pain sensations and the unmyelinated nerve fibers signal a more dull, aching type of pain. The tooth has few large myelinated nerve fibers that send information related to nonpainful sensations.

When the dentin is exposed, for example, by a cavity or a crack in a tooth, fluid in the dentinal tubules moves. This movement activates nerve fibers in the tooth and results in a sensation of pain. Similarly, procedures in the dental office, such as drilling, scraping, and drying the dentin will cause pain. Pain can also be caused by infection that inflames the tooth pulp.

The shot you get before the drilling begins is a local anesthetic that blocks nerve signals in the trigeminal nerve by stopping action potentials. Because nerve cells cannot send pain signals to your brain, you are pain-free and can relax as the dentist repairs your tooth.

? **?** **?** **?** **?**

What is synesthesia?

A

What if the number five sounded like a low-pitched tone or if something sweet felt round? This is how some people with synesthesia experience the world—as if one sense is joined with another. For example, a person with synesthesia may see a particular letter, shape, or number as having a specific color, sound, shape, or smell; a sound may have a particular color or taste. Any combination of the senses is possible. The word *synesthesia* comes from the Greek words, *syn* (together) and *aisthesis* (perception).

The combination of senses experienced by people with synesthesia is unique to each person and different people with synesthesia will almost always disagree on their perceptions. In other words, one person may think that the letter K is red, but another person might see K as blue. Synesthesia is not just people assigning one perception to another. Rather, people with synesthesia do not have to think about their perceptions—they just happen automatically. Also, the perceptions are the same over time. For example, the sound of middle C might always taste like coffee. These mixed perceptions often cause emotional reactions, especially pleasurable feelings, in people with synesthesia.

Estimates of the number of people with synesthesia range from 1 in 200 to 1 in 100,000. Many people with synesthesia probably don't even know they have it because they assume everyone perceives the world like they do. Women are three to eight times more likely to report that they have synesthesia, and more people who are left-handed say they have synesthesia. Synesthesia also appears to run in families, and there are some data that suggest a genetic link for the phenomenon.

The underlying brain mechanisms responsible for synesthesia are not well understood. Synesthesia may result from crossed wiring in the brain where neurons and their connections for one sensory system link to another sensory system. This crossed wiring likely involves the cerebral cortex and limbic areas of the brain.

What are hallucinations?

Hallucinations are sensations that seem real but are not. They don't seem like dreams, but they don't arise from external physical stimuli. They can be associated with any or all of the basic senses (vision, hearing, smell, taste, and touch), or they can be more abstract like a sense of being watched, a sense of purpose or imperative, or a feeling of spirituality or transcendence. Likewise they can be simple, like seeing shapes or hearing buzzing, or complex, like seeing people or hearing music.

Hallucinations can be the side effects of a number of different pathologies, including epilepsy, brain tumors, dementia, migraine, drug withdrawal, and vision loss. Of course, hallucinations can also be induced by various drugs, both natural and synthetic. Some of these substances have been used for thousands of years in religious contexts and in recreational contexts. The most famous of these is D-lysergic acid diethylamide (LSD), which was used heavily, at least by some people, in the 1960s and which the CIA famously (and horribly) investigated for use as a mind control agent.

There are currently no known medicinal uses for hallucinogenic compounds. LSD was originally made from a grain fungus called ergot, from which one of the first migraine drugs was also derived. Incidentally, ergot may have also played a role in the Salem witch trials and the French Revolution.

Hallucinations are internally sourced, but your brain can't tell. The hallucination is presented to your consciousness as if it were a real experience. Therefore, when people have hallucinations, they may or may not realize it. This inevitably leads to the question, "How can I know that anything I am experiencing is real, and not just a very complicated hallucination?" This is more a philosophical question than a scientific question. The scientific answer is, you can't. (If that is unsatisfying, go watch *The Matrix*.) Suffice it to say that if you see flashing lights because you are having a migraine headache, you are much more likely to recognize it is a hallucination than if you hear voices telling you that you are a god because you have schizophrenia.

Reference

National Institute on Drug Abuse, BrainFacts: Hallucinogens, January 2016, accessed February 9, 2016, http://www.drugabuse.gov/publications/drugfacts/hallucinogens.

What causes colorblindness?

To understand why some people cannot see colors, you have to understand why most people do. Light must pass through the cornea, pupil, and lens before it hits the retina at the back of the eye. The retina contains two types of special cells, called rod and cone photoreceptors, that are sensitive to light. Each eye has 5–6 million cones and 120–140 million rods. Rod photoreceptors respond to changes in light, movement, and shape, but they do not provide information about color. The three kinds of cone photoreceptors respond best to three different wavelengths of light. The relative differences in the responses of these three types of cones provide the brain with information about all of the colors we perceive.

People who are colorblind cannot tell some colors apart. Most people who are colorblind inherit the condition and are born without one type of cone receptor. The most common type of colorblindness involves red and green color vision. Even without one type of cone receptor, people can still see many colors, but they confuse certain ones. In the general population, about 8 percent of all men and 1 percent of all women are missing one type of cone. The inability to see any color is relatively rare, appearing in only 0.001 percent of the population.

A more rare condition called cerebral achromatopsia causes colorblindness even though cone receptors are normal. People with cerebral achromatopsia have damage to part of the visual cortex, and they see the world in shades of gray.

Color vision in dogs is similar to that in people with red-green colorblindness. Dogs have only two types of cones and likely see red as dark brown while green, yellow, and orange all look yellow.

What is umami?

Q

A

In 1908, Japanese scientist Kikunae Ikeda (1864–1936) isolated the amino acid glutamate from a seaweed called kombu. Ikeda proposed that glutamate provides foods with a unique taste that he named *umami*. Umami has been described as the savory taste experienced when you eat foods such as parmesan cheese, meat, tomatoes, seaweed, soy sauce, and yogurt.

By the late 1970s, anatomical and neurophysiological experiments confirmed that umami should join sweet, salty, bitter, and sour on the list of basic tastes. Just as the tongue has receptors on taste buds that respond to sweet, salty, bitter, and sour tastes, there are receptors that are activated by glutamate.

Have you ever wondered why MSG makes things taste so good? MSG stands for monosodium glutamate—this flavor enhancer binds to those glutamate (umami) receptors. Glutamate receptors have also been located in the stomach. These receptors send signals to the brain via the vagus nerve. The brain then responds by sending messages back to the digestive system to help process proteins.

Are there special places on the tongue for different tastes?

A

No matter how many times you see a "tongue map" in a textbook, it is still wrong. You may have seen a tongue map with sweet located on the front of the tongue, salty on the sides of the tongue, sour behind the area for salty, and bitter at the back of the tongue. Although these maps are nice and tidy, they are not based on scientific fact.

In 1901, German researcher David Pauli Hänig published his dissertation with illustrations of a fairly uniform taste map of the tongue for salty, sweet, bitter, and sour. In Hänig's publication, no part of the tongue was sensitive to only one particular taste; all parts of the tongue could detect all four different tastes. Some parts of the tongue were more sensitive than others. Through misinterpretation and translation errors of the dissertation, the tongue map was born and not challenged for many years.

We now know that taste buds are found not only on the tongue but also on the palate and cheeks. Taste buds respond to all tastes, not just one. What about umami, the fifth basic taste? It is never represented on any taste maps.

Still skeptical? Try it yourself. Put a bit of salt on the tip of your tongue, the classic "sweet area" and see if you can taste it. Or try a bit of sugar on the side of your tongue, the "sour zone."

Reference

Hänig, D.P., *Zur psychophysik des geschmackssinnes* (Engelmann, 1901).

? ? ? ? ?

Can listening to loud music damage your hearing?

A

Let's hope you are not blasting music as you are reading this because, yes, loud sounds can damage your hearing. Sound is carried by changes in air pressure that vibrate the eardrum (tympanic membrane). Humans can hear sounds waves that have frequencies between 20 and 20,000 Hz. The vibration of the eardrum moves three small bones in the ear (malleus, incus, stapes) that pass the vibrations to hair cells inside the cochlea of the inner ear. The cilia (the hair) of the hair cells help generate electrical signals in the auditory nerve that are sent to the brain.

Our ears are very sensitive to loud noises. These noises can damage and even kill hair cells. The death of hair cells means that information about sound will not be sent to the brain. According to the U.S. Centers for Disease Control and Prevention:

- Four million workers are exposed to damaging noise at work every day.

- Ten million people in the United States have a noise-related hearing loss.

- Twenty-two million workers are exposed to potentially damaging noise each year.

The intensity (loudness) and duration (time) of a sound contribute to the amount of hearing damage. For example, a one-time exposure to the loud bang of a firecracker or longtime exposure to heavy car traffic can cause hearing damage.

But don't despair. You can prevent hearing problems while still enjoying your favorite music. Just by reading this information, you now know the risks of loud music. You know that you need to turn the volume down and limit the time you are exposed to loud sounds. If you are at a concert, don't stand in front of the loudspeakers; find a quieter place where you can still enjoy the music or wear ear plugs to protect your ears. If you think you are having trouble with your hearing, see a health care provider who can test your hearing.

Reference

Centers for Disease Control and Prevention, Noise and hearing loss prevention, January 25, 2016, accessed February 15, 2016, http://www.cdc.gov/niosh/topics/noise/.

Why can't I tickle myself?

A

You might be surprised to know that this is a serious scientific question. "Why Can't You Tickle Yourself?" was the title of a scientific paper published in the journal *NeuroReport*. Although the ability to tickle oneself is not a significant health concern, the mechanisms involved with this phenomenon say something about how we pay attention to events we cause ourselves compared to when the events are caused by someone or something else.

Scientists from the University College London constructed a robotic arm with a piece of foam that they used to touch the right hand of their subjects. In one condition, the subjects moved their left hand that controlled the robotic arm; in another condition, the robot moved its own arm to touch the subjects' hands. People found that when they moved the robotic arm with their left hands, the sensation they felt was less tickly and less pleasant than when the same touch was generated by the robot. Also, as the delay and direction between a subject's left hand and the resulting touch were increased, the touch was rated as more tickly.

These data suggest that being able to predict the nature of a stimulus (a tickly touch) leads to a less tickly sensation. In other words, when you know when and how you will be tickled, you find it less ticklish.

Further investigation using brain imaging points to the cerebellum as an important area responsible for this perception. Brain activity in the cerebellum is reduced when people try to tickle themselves. Similar movements without an actual tickle do not change cerebellar activity. More generally, the cerebellum helps predict what something will feel like based on the movement that causes the sensation. Why is this a useful thing to do? Possibly because it helps your brain tell the difference between internally generated sensations and externally generated sensations.

References

Blakemore, S.-J., Wolpert, D., and Frith, C., Why can't you tickle yourself?, *NeuroReport*, 11:11–16, 2000.

Blakemore, S.-J., Wolpert, D.M., and Frith, C.D., Central cancellation of self-produced tickle sensation, *Nature Neuroscience*, 1:635–40, 1998.

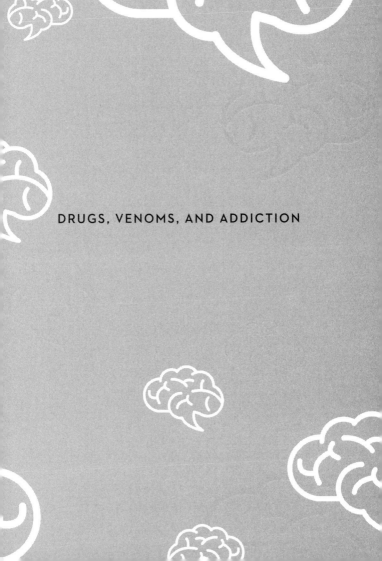

DRUGS, VENOMS, AND ADDICTION

Does alcohol kill brain cells?

A

People who are drunk certainly seem as if they have killed more than just a few brain cells. But moderate consumption of alcohol does not cause nerve cells to die directly.

Ethanol (ethyl alcohol), the type of alcohol found in beer, wine, and spirits, is quickly absorbed by the stomach and small intestine into the blood stream. Once in the blood stream, alcohol makes its way throughout the body. Because it is soluble in fat and water, alcohol has an easy time crossing the blood-brain barrier, where it can act on brain cells.

Once in the brain, alcohol targets several different neurotransmitter systems. GABA is a neurotransmitter that most often decreases (inhibits) neural activity after it binds to a receptor. Alcohol increases the activity at GABA receptors so it reduces brain activity. Glutamate, on the other hand, is a neurotransmitter that increases (excites) neural activity. Alcohol reduces the activity at glutamate receptors. The combined effects of alcohol on GABA and glutamate neurotransmitter systems lead to an overall reduction in brain activity. The signs of intoxication such as slurred speech, memory problems, difficulty balancing, and nausea depend on which parts of the brain are affected. A third neurotransmitter system, one that uses dopamine, is also activated by alcohol. Dopamine levels in the brain are boosted while the amount of alcohol in the blood increases. The rush of dopamine gives rise to the pleasurable feelings associated with drinking alcohol. Interaction with the dopaminergic system is also what causes alcohol addiction.

The changes in GABA, glutamate, or dopamine neurotransmitter systems caused by moderate alcohol consumption do not result in cell death. Chronic, long-term drinking, however, can lead to serious health problems, including dependence and addiction. Alcoholism can lead to a failure of the digestive system to absorb vitamin B1 (thiamine) properly. A vitamin B1 deficiency can cause Wernicke's encephalopathy, a disorder characterized by memory, thought, balance, and movement problems. More severe lack of vitamin B1 can lead to Korsakoff's syndrome, with more serious memory and cognitive problems. Neurons in various parts of the brain including the thalamus, cortex, cerebellum, and hypothalamus are destroyed as a consequence of long-term vitamin B1 deficiency. So when consumed in excess, alcohol can indirectly kill brain cells.

References

Harper, C., The neuropathology of alcohol-specific brain damage, or does alcohol damage the brain?, *Journal of Neuropathology and Experimental Neurology*, 57:101–10, 1998.

Harper, C., Dixon, G., Sheedy, D., and Garrick, T., Neuropathological alterations in alcoholic brains. Studies arising from the New South Wales Tissue Resource Centre, *Progress in Neuro-Psychopharmacology and Biological Psychiatry*, 27:951–61, 2003.

Why do sleeping pills make me sleepy?

A

Many people turn to medication to help them fall asleep and stay asleep. The drugs used to fight insomnia can be divided into a few distinct groups based on how they work.

Barbiturates, first developed in the late 1800s, were introduced into medical practice in the early 1900s. Pentobarbital, phenobarbital, secobarbital, and amobarbital are all types of barbiturates. Barbiturates act on GABA and glutamate neurotransmitter systems in the brain. When GABA binds to receptors on neurons, neural activity is reduced. When glutamate binds to its receptors, it has an opposite effect: increased neural activity. Barbiturates work to enhance the GABA system and inhibit the glutamate system, so the combined overall effect is reduced brain activity.

At low doses, barbiturates cause sedation, reduced heart rate, reduced respiration, and sleep. Higher doses of barbiturates can cause anesthesia and an overdose can send someone into a coma because of the drug's effect on respiratory areas of the brain. Because barbiturates have a high risk of abuse, tolerance, and addiction, these drugs were replaced by a class of drugs called the benzodiazepines.

Benzodiazepines were introduced in the 1960s as alternatives to barbiturate drugs. Common benzodiazepines are Xanax (alprazolam), Klonopin (clonazepam), Valium (diazepam), ProSom (estazolam), Dalmane (flurazepam), Ativan (lorazepam), Restoril (temazepam), and Halcion (triazolam). Like barbiturates, benzodiazepines also work to enhance the GABA neurotransmitter system. Benzodiazepines are safer than barbiturates because they do not cause breathing problems. The key to the action of benzodiazepine drugs is that these chemicals bind to a special place on the GABA receptor, and few of these binding sites are found on neurons that control respiration.

Benzodiazepines are not perfect sleeping pills because they still can cause movement problems and trouble with learning and memory. These problems lead to the development of new sleep drugs such as Ambien (zolpidem) and Lunesta (eszopiclone). Although these medications have different chemical structures than do benzodiazepines and barbiturates, they still act on GABA receptors. These newer drugs are more selective because they bind primarily to GABA receptors on neurons involved with sleep, but they still have a number of adverse side effects such as amnesia and sleep eating.

All sleeping pills should be taken right before bedtime. Because the purpose of the medication is to help people fall asleep and stay asleep, anyone who has taken a sleeping pill should not drive. Sleeping pills can become habit-forming and you shouldn't take them all the time. Medications can help you get some shut-eye in the short term, but in the long run you're much better off addressing the reasons you can't sleep than continuing to take drugs.

How does coffee wake me up?

A

Coffee is a complex combination of hundreds of different chemicals that gives the drink its characteristic flavor and aroma. The chemical that provides the wake-up call in the morning? Caffeine!

Caffeine belongs to a class of compounds called xanthines and has multiple actions on the body. Caffeine works as a stimulant because it blocks the effects of the neurotransmitter adenosine in the brain. When adenosine normally binds to receptors in the brain, neural activity slows down. This reduction in activity causes a person to feel tired, drowsy, and sleepy. Caffeine blocks adenosine receptors and thus has the opposite effect: increased neural activity and stimulation.

The brain is not the only place where caffeine acts because adenosine receptors are found in other places in the body. For example, caffeine increases heart rate, blood pressure, and respiratory rate. That need to pee after your morning coffee? That's because caffeine blocks adenosine receptors on the kidneys (and in the brain). Blood vessels also constrict when adenosine receptors are blocked by caffeine. Headache medication often includes caffeine to take advantage of this effect on blood vessels in the brain.

The neurological effects of caffeine do not stop with adenosine. Caffeine can increase the release and may slow the reuptake of dopamine, another neurotransmitter. Caffeine's effect on the brain's dopamine system may contribute to the pleasurable feelings associated with the morning cup of joe.

Coffee is not the only source of caffeine: tea, some sodas, energy drinks, and chocolate all contain varying amounts of the chemical. A cup of drip coffee has 60-150 mg of caffeine and a cup of tea contains 25-61 mg of caffeine. A 16 oz. energy drink can have more than 200 mg of caffeine, whereas a can of cola soda has about 50 mg of caffeine. Even chocolate contains a small amount of caffeine, about 10 mg in a chocolate candy bar. Regardless of how you get your caffeine, it will have the same effect on your body and brain.

Reference

Center for Science in the Public Interest, Caffeine content of food & drugs, accessed January 10, 2016, http://www.cspinet.org/new/cafchart.htm.

Is there a pill to cure alcohol addiction?

Many people are able to enjoy alcohol in moderate quantities and may even enjoy some health benefits from doing so. For others, alcohol becomes a controlling demon, the need for which supersedes all else. In this population, the pursuit of alcohol often wreaks havoc on employment, interpersonal relationships, health, and just about every other measure of well-being.

Alcohol addiction is very egalitarian, destroying the lives of senators, movie stars, and ordinary citizens alike. For alcoholics, some combination of genetic predisposition and environmental factors leads to addiction, which is a chronic brain disease characterized by compulsive drug seeking and use. Even when people addicted to alcohol recognize that they have a problem and earnestly desire change, it is difficult and sometimes impossible to quit. This is because drug abuse recalibrates the reward circuits in the brain, making them less responsive to ordinary life and more geared towards the drug. Wouldn't it be nice if medicine could do something about that?

It turns out that doctors would really like to have a fix for alcoholism, not the least because they also suffer from the disease. Case in point: Olivier Ameisen, a French cardiologist who in 2005 published his own case study of alcoholism and the drug treatment that "cured" him—baclofen. Ameisen found that taking regular, high doses of baclofen eliminated his cravings for alcohol and allowed him to quit drinking effortlessly. Baclofen is not a new drug: it has been on the market for many years as a muscle relaxant commonly used to treat spasticity. Ameisen was also not the first person to realize that baclofen could be used to reduce addictive cravings because it had been shown previously to reduce symptoms in cocaine addicts. What Ameisen did do was popularize the use of baclofen as a treatment for alcoholism. Since his book was published, the drug has been increasingly used "off-label" to treat addiction.

Does baclofen work? It is somewhat difficult to say, because there have not yet been any large studies, and the smaller studies have had conflicting results. There are many empirical results to suggest that baclofen does reduce or eliminate cravings for alcohol. As a bonus, it is does not have any inherent psychotropic or addictive qualities, is generally well tolerated, effectively reduces anxiety, and can be used to treat the symptoms of alcohol withdrawal. Is it a magic bullet? Probably not. For one thing, for it to be effective, you have to take it in perpetuity, and long-term compliance is always an issue. In addition, even though the side effects are usually minor, there are side effects, especially for the high dosages required to eliminate cravings. These side effects can include insomnia, nausea, dizziness, muscle pain, motor instability, visual problems, decreased libido, sweats, cramps, sleepwalking, and ringing in the ears. You can see why long-term compliance might be a problem. Baclofen also treats only the physical cravings of addiction: it does not remove the social component of drinking and does not address any underlying issues that might have sparked the drinking in the first place.

Even when people are able to completely control their cravings with baclofen, they sometimes go off the drug so they can go on a weekend bender. It is possible to acknowledge that an alcohol addiction is ruining your life and still want to be drunk. That means that baclofen may be a good tool. But it isn't enough. The issues that cause addiction are complex, and so must be the remedy.

References

de Beaurepaire, R., The use of very high-doses of baclofen for the treatment of alcohol-dependence: a case series, *Frontiers in Psychiatry*, 5, 2014, doi:10.3389/fpsyt.2014.00143.

Gorsane, M.-A., Kebir, O., Hache, G., Blecha, L., Aubin, H.J., Reynaud, M., and Benyamina, A., Is baclofen a revolutionary medication in alcohol addiction management? Review and recent updates, Substance Abuse 33:336-349, 2012.

National Institute on Drug Abuse, DrugFacts: Understanding drug abuse and addiction, November 2012, accessed January 12, 2016, http://www.drugabuse.gov/publications/drugfacts/understanding-drug-abuse-addiction.

Is marijuana addictive?

An estimated 9 percent of people who try marijuana will develop a dependency. This number goes up to about 17 percent if for those who start using while they are young. For daily users the odds are about 25–50 percent. So, yes, marijuana is addictive. It is not as addictive as, say, heroin, nor does it have the same potential for harm.

The withdrawal symptoms from marijuana last up to two weeks and are fairly mild. They include irritability, mood swings, sleep disturbances, decreased appetite, cravings, and restlessness. This doesn't make for a fun fortnight. But compared to the withdrawal symptoms for alcohol, which are potentially fatal and include hallucinations, seizures, and delirium tremens, they aren't so severe.

Does this mean that marijuana is harmless? No. Like alcohol, marijuana can have profound and lasting effects on the brain, especially on the developing brain (and note that the brain develops all the way through teenage years). The most well-documented effect of heavy use is memory impairment, but regular use of marijuana during teenage

years can also shave off an average of eight IQ points. Those IQ points are not recoverable. Marijuana use has also been linked to schizophrenia in people who have a genetic predisposition to the disease. In men, it has been linked to an aggressive form of testicular cancer.

Is it better to use marijuana than methamphetamines? Yes, but you really shouldn't use either one. If you really need to get high, try going for a run and eating some spicy food (in that order). Even if you don't achieve the euphoric high that some people experience, your jeans will fit better and your sinuses will be clear.

Reference

National Institute on Drug Abuse, Is marijuana addictive?, September 2015, accessed January 28, 2016, http://www.drugabuse.gov/publications/research-reports/marijuana/marijuana-addictive.

How does LSD cause hallucinations?

Chemist Albert Hofmann (1906–2008) had an unusual experience on his way home from work one day in 1943. Soon after he drank a solution that he mixed in his lab, he started to feel dizzy and anxious, so he got on his bike and made his way home. During the trip home, Hofmann's vision blurred and his sense of his body changed. Once home, he experienced threatening, frightening visual hallucinations and felt as if he was outside his body. This was because he had just downed a concoction of lysergic acid diethylamide (LSD).

In addition to causing visual hallucinations, LSD can produce auditory and somatosensory hallucinations and distort perceptions of time and space. These hallucinations can be accompanied by changes in mood (panic, confusion, happiness, or sadness), faster heart rate, higher blood pressure, chills, and weakness.

The underlying brain mechanisms responsible for the effects of LSD have to do with the similarity of this drug to the neurotransmitter serotonin. To a receptor on a neuron, LSD and serotonin look the same. In fact, neurons are more likely to be activated by molecules of LSD than they are to serotonin. Therefore, brain pathways that use serotonin for signaling, such as those used for perception and emotions, will become overexcited. This increased activation will cause a person to hear, see, smell, taste, or feel things that are not actually present.

LSD is illegal to possess, manufacture, and buy in the United States and other countries. So in addition to a bad trip filled with terrifying sights and sounds, LSD users may also suffer a bad trip to jail. In fact, the U.S. Controlled Substances Act states that the punishment for being caught with 10 grams or more of LSD is a jail sentence of at least 10 years.

References

Hofmann, A., *LSD, My Problem Child* (New York: McGraw-Hill, 1980).

U.S. Food and Drug Administration, Controlled Substances Act, June 11, 2009, accessed February 26, 2016; http://www.fda.gov/regulatoryinformation/legislation/ucm148726.htm.

Is ecstasy ("Molly") dangerous?

A

The drug known as ecstasy or Molly has the long chemical name 3,4-methylenedioxy-methamphetamine, abbreviated MDMA. Although MDMA was synthesized in the laboratory more than 100 years ago, it did not gain in popularity until the late 1980s. Users of MDMA report intense feelings of joy, heightened emotions, increased energy, and an altered sense of time and their environment.

In the brain, MDMA causes increases in the levels of dopamine, norepinephrine, and serotonin; these changes are responsible for the high that a user experiences. Research into how MDMA may be toxic to the brain got off to a rough start when one laboratory reported that MDMA damaged neurons that contained dopamine. On further examination of their procedures, the researchers reported that the drug bottle was mislabeled and they had actually administered methamphetamine instead of MDMA to their experimental animals. Further investigations by several laboratories have found that MDMA damages neurons that contain serotonin, especially in the hippocampus and frontal cortex of the brain. Some studies indicate that MDMA users have problems with their memory and other cognitive functions, while other studies have failed to confirm these findings.

MDMA users often suffer from cramps, nausea, teeth clenching, excessive sweating, and visual problems. These side effects may be the least of the problems. MDMA-induced increases in body temperature and brain swelling can be lethal, and multiple uses of MDMA can cause inflammation of the heart and irregular heart rhythms.

In addition to damaging the nervous system, MDMA may harm the cardiovascular system.

MDMA shares another problem common to many other illegal drugs: it may be adulterated with other chemicals and the dose from one pill to the next may be different. Therefore, people who think they are taking ecstasy may actually be ingesting any number of other drugs such as cocaine, bath salts, or amphetamines, and they may not know how much of the drug they swallow.

References

Badon, L.A., Hicks, A., Lord, K., Ogden, B.A., Meleg-Smith, S., and Varner, K.J. Changes in cardiovascular responsiveness and cardiotoxicity elicited during binge administration of ecstasy, *Journal of Pharmacology and Experimental Therapeutics*, 302:898–907, 2002.

Parrott, A.C., Human psychobiology of MDMA or "Ecstasy": an overview of 25 years of empirical research, *Human Psychopharmacology*, 28:289–307, 2013.

Ricaurte, G.A., Yuan, J., Hatzidimitriou, G., Cord, B.J., and McCann, U.D., Retraction, *Science*, 301:1479b, 2003.

Schifano, F., Oyefeso, A., Corkery, J., Cobain, K., Jambert-Gray, R., Martinotti, G., and Ghodse, A.H. Death rates from ecstasy (MDMA, MDA) and polydrug use in England and Wales 1996–002. *Human Psychopharmacology*, 18:519–24, 2003.

Can some drugs turn people into zombies?

A

If you think of a "zombie" as a person who has died and then has been reanimated to terrorize and munch on the flesh of the living, then, no. There are no drugs capable of creating a supernatural being.

Reports of Haitian voodoo doctors turning people into zombies should be viewed skeptically. In the 1980s, Wade Davis published an article and a book that detailed how Haitian voodoo priests used tetrodotoxin to create the living dead. Tetrodotoxin is a potent neurotoxin found in the pufferfish, blue ringed-octopus, and some salamanders, newts, and frogs. This neurotoxin blocks sodium channels on nerve cells, thereby preventing the conduction of electrical messages. Tetrodotoxin can paralyze muscles, interfere with breathing, and in a few cases every year, kill. A potential zombie may also be given a drug from the jimson weed to cause hallucinations and confusion. Davis claimed that tetrodotoxin was the primary ingredient in zombie powder, responsible for creating a zombie-like state, but researchers who tested samples of the powder failed to find significant amounts of the chemical. Furthermore, someone poisoned by tetrodotoxin would be unable to move and would not have the strength to stagger around looking for a victim to bite.

Other drugs linked to the creation of zombies include desomorphine, bath salts, and phencyclidine. Desomorphine, also called krokodil, is a highly addictive painkiller that can rot teeth and eat away at the flesh of users, resulting in the look of a typical zombie. Bath salts and phencyclidine (PCP or angel dust) may result in the manic, aggressive behavior of a zombie. Bath salts are synthetic chemicals similar to cocaine, amphetamines, and ecstasy that boost brain levels of the neurotransmitters dopamine and serotonin. Initial feelings of pleasure and euphoria created by bath salts can be replaced with paranoia, fear, and stress along with high blood pressure and heart rate. Add sleep deprivation, increased anxiety, an increased tolerance to pain, and threatening hallucinations to the profile and you have a person who may act out aggressively. People who use PCP may also seem aggressive and are sometimes viewed as violent criminals for the way they act. PCP affects multiple neurotransmitter systems and can cause hallucinations and a break from reality.

All of these drugs alter behavior by changing brain chemistry. None of them will wake the dead, however, so you can put your survival strategies for the zombie apocalypse on the back burner.

References

Davis, W., The ethnobiology of the Haitian zombie, *Journal of Ethnopharmacology*, 9:85-104, 1983.

Drug Enforcement Administration, Desomorphine, accessed January 22, 2016, http://www.deadiversion.usdoj.gov/drug_chem_info/desomorphine.pdf.

Yasumoto, T., and Kao, C.Y., Tetrodotoxin and the Haitian zombie, *Toxicon*, 24:747-49, 1986.

Why do cats like catnip?

One minute your cat is a friendly fur ball; then, after one or two whiffs of catnip, it turns into a frenzied feline fidgeting all over the floor.

Catnip (*Nepeta cataria* and other *Nepeta* species), a plant from the mint family, has a seemingly magical power over many cats that sends them into a state of ecstasy. The plant affects 50-80 percent of all cats, and the sensitivity to the smell appears to be inherited. Interestingly, kittens do not respond to catnip until they are three to six months old.

The special "sauce" inside catnip's leaves, stems, and seeds is a volatile oil called nepetalactone. Nepetalactone probably mimics natural pheromones that cats encounter as they prowl their neighborhoods. After nepetalactone binds to receptors in a cat's nose, the olfactory nerve sends signals to brain areas including the amygdala and hypothalamus. These signals cause the cat to sniff, lick, rub themselves, roll around, and generally act as if they are having a great time. This behavior usually lasts for about 10 minutes and then the cat loses interest.

Catnip does not have similar effects on people, although some people brew catnip tea to relax or treat insomnia and headaches. You might enjoy a cup of catnip tea, but it won't get you high.

Reference

Hart, B.L., and Leedy, M.G., Analysis of the catnip reaction: mediation by olfactory system, not vomeronasal organ, *Behavioral and Neural Biology*, 44:38-46, 1985.

Can bee stings cure multiple sclerosis?

A

Multiple sclerosis (MS) is a disease characterized by deterioration of the insulating sheath (myelin) that surrounds axons in the central nervous system. The insulation is there to increase the speed of communication between neurons, and its absence causes communication breakdown. This results in a number of undesirable outcomes, including visual impairment, muscle weakness and paralysis, abnormal sensations, pain, speech impediments, hearing loss, and cognitive impairment. MS can be progressive, or it can be relapsing and remitting. No one really knows what causes MS, but many scientists believe it is an autoimmune disease. Some of the symptoms can be treated, but unfortunately, there is no known cure. As a result, many people who suffer from MS have turned to alternative medicine. One of the more highly publicized alternative medicine treatments is bee sting, or bee venom therapy. Bee stings have been traditionally used in Eastern medicine to treat pain and inflammation. Some scientific studies have shown that in fact, bee venom can have an anti-inflammatory effect in animal models. MS is an inflammatory disease, so there was hope that bee stings could ameliorate some of the symptoms. Indeed, some individuals have reported dramatic, almost magical reversals of their disease state following bee sting therapy. In this therapy a patient is exposed to the stings of many, sometimes hundreds, of bees at a time (this is less therapeutic for the bees because they die after they sting). For many of us this might be a nightmare scenario, but for some the pain of the stings pales in comparison to the pain of MS. Better still, aside from the unpleasantness of being regularly stung by lots of bees, the therapy is somewhat safe as long as you aren't allergic to bee stings.

Unfortunately, a clinical trial found that 24 weeks of bee sting therapy was no more effective than a placebo at reducing brain lesions, preventing relapses, or improving disability, fatigue, and quality of life in patients with relapsing MS. In this study the maximum number of stings delivered at a time was capped at 20. This number of bee stings is substantially fewer than some people use, so the dose in that study may have been insufficient. However, the treatment of MS with bee stings remains scientifically unsupported. Incidentally, bee venom has also been reported to have an antitumor effect in some animal models, but it has not proven to be an effective cancer treatment in any human cancers. On the other hand, honey, another product from bees is delicious and is demonstrably useful in wound care and as a cough suppressant.

Other animal venoms, including those from sea anemone, scorpion, and snake, have been investigated as treatments for MS. Unlike bee venom, the potentially therapeutic nature of these venoms is not anti-inflammatory but has to do with their ion channel blocking properties. This is actually what makes these venoms, well, venomous. When used carefully, they may improve the conductive capacity of demyelinated neurons and thus reduce the symptoms of MS. Results are very preliminary, however, and handling poisonous animals is ill-advised.

References

Mirshafiey, A., Venom therapy in multiple sclerosis, *Neuropharmacology*, 53:353–61, 2007.

Wesselius, T., Heersema, D.J., Mostert, J.P., Heerings, M., Admiraal-Behloul, F., Talebian, A., van Buchem, M.A., and De Keyser, J., A randomized crossover study of bee sting therapy for multiple sclerosis, *Neurology*, 65:1764–68, 2005.

What animals have venom that attacks the nervous system?

A

Pain, itching, headache, paralysis, death: just a few of the unpleasant consequences of encountering the neurotoxic venom from a variety of creatures. The Earth is filled with many animals, big and small, with an arsenal of chemical weapons that target the nervous system of other animals. These neurotoxins help animals defend themselves from predators or capture a tasty meal.

Animal neurotoxins affect the way nerve cells transmit information and can alter neuronal activity in several ways. Neurotoxic venoms can change the way neurotransmitters are released from axon terminals. For example, latrotoxin from the venom of the black widow spider, causes a massive release of the neurotransmitter acetylcholine. Other neurotoxins, such as crotoxin from the South American rattlesnake, inhibit the release of acetylcholine. Because acetylcholine is involved with the contraction of muscles, alteration in the way muscles receive these chemical signals can affect movement and breathing.

Neurotoxic venoms can also act on receptors that bind neurotransmitters. Receptors that use acetylcholine are blocked by the venoms of several snakes, and receptors that use the neurotransmitter glutamate are blocked by the venoms of the predaceous wasp and Joro spider. Instead of blocking receptors, domoic acid from the blue mussel excites glutamate receptors.

Many neurotoxins affect neurotransmission by altering the opening and closing of potassium and sodium ion channels. The most well-known neurotoxin is tetrodotoxin, a chemical found in the skin and internal organs of the pufferfish. Tetrodotoxin blocks sodium channels and therefore prevents the generation of action potentials. Most scorpion venoms block potassium channels and, like the venom of the pufferfish, blue-ringed octopus venom blocks sodium ion channels. Calcium channels are blocked by the venoms found in the funnel web spider and black mamba snake.

Although the animal neurotoxins found in snakes, spiders, and scorpions are the most well known, neurotoxins are also found in other animals. The skin of poison arrow frogs, for example, contains the extremely toxic batrachotoxin, which prevents sodium channels from closing. Marine snails use conotoxin to block calcium channels. There is even a bird that has a neurotoxin: the pitohui contains homobatrachotoxin, a chemical that activates sodium channels.

Animals use neurotoxins for predation and protection, but humans have also found these powerful chemicals to be useful to treat disease. For example, neurotoxins are used to create antivenoms that have saved the lives of many people who have been bitten by snakes and spiders. Conotoxin, the venom used by marine snails, has made medicine to alleviate chronic pain. Future researchers will likely tap into nature's treasure chest of neurotoxins for drugs to treat many other neurological problems.

Appendix 4 lists many of the neurotoxins found in animals.

POPULAR CULTURE

Why do we use only 10 percent of our brain?

A

If you could use 100 percent of your brain, then you would be able to move objects without touching them, memorize pages of information with a single glance, know the future, levitate, and control time. Right? Wrong! There is no evidence we are using only 10 percent of our brain, and there is no reason any amount of brain usage would result in a violation of the laws of nature.

When people talk about brain usage, they most often refer to the amount of brain tissue that is necessary to function normally. The often quoted figure of 10 percent implies that 90 percent of the brain is sitting in our head doing nothing. Taken further, this assertion suggests that if 90 percent of the brain is removed, a person would be just fine. Such statements are ridiculous and not based on anything we know about brain function. In fact, all evidence points to the fact that we use our entire brain.

Functional brain imaging methods (e.g., functional magnetic resonance imaging or positron emission tomography) show that the entire brain is active. Certainly different tasks cause more or less activity in different parts of the brain, but there are no areas of the brain without brain activity. Similarly, electrical recordings from the brain using electroencephalography (EEG) show activity at all locations. The brain is active even when a person is sleeping.

If 90 percent of the brain was not being used, then removal or injury to large areas of the brain should not result in significant problems. Unfortunately, damage to small areas of the brain, for example, after a stroke or head injury, can have catastrophic consequences on the ability to move, speak, hear, or feel. Diseases of the nervous system also may affect only a limited amount of brain tissue but can alter a person's behavior significantly.

The brain's neurons do not like to sit idle, and research shows that if they are not active, they will wither and die. This is especially apparent when the brain is developing: if neurons are prevented from receiving sensory signals, connections between neurons responsible for that sense will be lost and the pathways that depend on those neurons will deteriorate.

If we are already using 100 percent of the brain, how can we learn new information and skills? Each time we learn something, it is not because we are tapping into some previously dormant area of the brain. Rather, we are forming new or stronger connections between neurons and creating new neural circuits. It is true that some children can recover from surgery that removes a large portion of the brain (but not 90 percent). These children are still using 100 percent of their existing brain, and because of the great flexibility of the young brain to rewire itself, they can largely recover. Our potential to learn is huge, but it is not because 90 percent of the brain is waiting for something to do.

? **?** **?** **?** **?**
Q

How do you become a brain researcher?

A

The road to a career in neuroscience can take many twists and turns and is sometimes filled with challenges and obstacles. The most important characteristics of a brain researcher are a desire to learn and a curiosity about the natural world.

Neuroscientists start with a good educational foundation. Rarely does a child decide on a career before high school. Regardless of when the decision to become a brain researcher is made, a potential neuroscientist usually takes a variety of classes in high school to prepare for a successful college experience. This means that students take biology, chemistry, and math classes as well as language and social science classes. Once in college, students can choose from a variety of majors. Not all colleges and universities offer undergraduate degrees in neuroscience or neurobiology. Students who major in other fields such as psychology, biology, chemistry, physiology, physics, or engineering can take courses to learn about neuroscience. Undergraduates who receive their degrees from nonscience departments can also pursue their interest in neuroscience if they take classes to prepare themselves for future study.

The perfect time for testing out a career in neuroscience is when students are in college. Undergraduate students can work in research labs to gain experience. The pay may not be great, but the people students meet and the knowledge they gain are valuable because these experiences can be used as stepping steps to the next level.

After graduating from college, students are ready for the next phase of the journey: graduate school or a health profession school (medical or dental school). Graduate students earn a Ph.D. after performing research on a particular topic in a neuroscience, biology, psychology, bioengineering, pharmacology, or other department. Medical and dental students are trained to diagnose and treat patients and may work in research laboratories.

Graduate school usually takes four to seven years to complete a Ph.D. program. Freshly minted Ph.D.s can get teaching and research jobs in companies and at universities, but competition for these positions is strong. Instead of going directly to industry or academia, many new Ph.D.s and some M.D.s take postdoctoral fellowships for additional training. During a postdoc, neuroscientists work in a lab to learn new procedures or investigate a new topic of interest. Short postdoctoral fellowships last only one year; some people remain as postdocs for five years or more before they move on to a position at a company or in a government lab or join the ranks as a professor at a university.

Does the brain work like a computer?

A

The brain has always been compared to the most advanced technology available at the time. For example, the brain was once thought to work like a water clock; later it was said to be like a telephone switchboard. These days it is popular to compare the brain to a computer.

There are many similarities and differences in the way brains and computers work. Both computers and brains need energy for power. Computers need a source of electricity to work, but brains must have oxygen, sugar, and minerals to survive. Electrical signals are also used by computers and brains to send messages. However, the electricity used by computers involves the direct flow of electrons; nerve cells generate electrical signals by differences in charged particles (ions) inside and outside of cells. Messages travel much faster within the wires of a computer than they do through the axons of a nerve cell. Nerve cells also use chemicals to send messages from one nerve cell to the next.

Computers direct their messages using switches that are either on or off. Neurons also send messages that are on or off in the form of action potentials. Action potentials are sent completely or not at all, a bit like a switch turning on and off. But neurons are more than binary switches because they are always working, even when action potentials are not being sent. Unlike a brain, a computer can be turned on and off by flicking a switch. The brain has no "off" button: it works nonstop, even when you are asleep.

A major similarity between computers and brains is that they both store information, and this ability to remember can grow. A computer's memory increases when computer chips are installed or when information is written to a disk drive or other storage device. The brain stores memories in neural circuits that can get more efficient when connections between nerve cells are strengthened. Unlike computer memory, which is essentially a perfect copy of what was entered, memories encoded into the brain are subject to change over time and situation.

In addition to having a memory, computers and brains can learn. Computers can run new programs and perform many jobs at the same time. Although brains multitask, too, for example by controlling heart rate, breathing, and moving simultaneously, they are not very good at performing multiple complex mental tasks simultaneously. Try saying the alphabet and adding two numbers at the same time.

Brains and computers don't last forever, and they are both subject to damage. Computers have hard drives that fail, they can be infected with viruses, and keyboards and monitors break. Replacement parts easily repair damage done to a computer. Unfortunately, the brain is not so easy to fix after it is damaged. You cannot pick up a new amygdala or a spare thalamus at your local store, and research using transplanted neurons has had only limited success. The brain does have some ability to repair itself by rewiring connections or using other pathways for those that are damaged.

? ? ? ? ?

Why do songs get stuck in my head?

You most likely have had one and wanted to get rid of it, but you may not have known what this phenomenon is called. It is an "earworm," a song or tune that keeps repeating over and over in your head and just won't leave.

Dr. James Kellaris, a professor in the College of Business at the University of Cincinnati, coined the term *earworm* after the German word *ohrwurm* and has studied how earworms burrow into your mind. Kellaris found that 97.9 percent of the people he surveyed had an earworm at some time, and women and musicians seem to get songs stuck in their heads more often than do men and nonmusicians. Typical earworms are songs with lyrics and advertising jingles. Songs reported to cause the most earworms include the "Baby Back Ribs" jingle from Chili's restaurant, "Who Let the Dogs Out?" performed by the Baha Men, and "We Will Rock You" by Queen. (Apologies if you now have these tunes implanted in your brain.)

How earworms get lodged in our brains is not known. The brain may try to fill in gaps in a song's rhythm, a bit like an auditory blind spot. If a rhythm is catchy, the brain may become infected with an earworm when the song continues to play in the mind even after it has ended.

There is no perfect cure to eliminate an earworm. Remedies that work for some people include listening to another song, singing another song, playing a different song on an instrument, and changing activities. Chewing gum can also reduce the number of songs that stick in your head. Presumably, the act of chewing blocks the ability to recollect musical and verbal memories.

References

Beaman, C.P., Powell, K., and Rapley, E., Want to block earworms from conscious awareness? B(u)y gum!, *Quarterly Journal of Experimental Psychology*, 68:1049–57, 2015.

Kellaris, J.J., Dissecting earworms: further evidence on the "song-stuck-in-your-head" phenomenon, in Christine Page and Steve Posavac, eds., Proceedings of the Society for Consumer Psychology Winter 2003 Conference, New Orleans, LA, American Psychological Society, 220–22, 2003.

Why do we yawn and why are yawns contagious?

A

Yawns: they last about six seconds; the first ones happen about 11 weeks after conception; they become contagious when people are one or two years old; the hypothalamus is involved in their generation; and no one is really sure why we have them.

People assume that yawns are caused by boredom, and there is some truth to this observation. Looking at a boring TV test pattern does cause more yawns than watching music videos. But that's not the whole story. Many athletes yawn just prior to taking part in a race or game, a time that should not be boring at all.

A common theory for the cause of yawning is a buildup of carbon dioxide and lack of oxygen. After all, a yawn involves a deep inhalation bringing air into the lungs.

Unfortunately, this idea is not supported by experimental evidence. In the late 1980s, Dr. Robert Provine counted the number of yawns in people who breathed in 100 percent oxygen, a mixture of 3 percent carbon dioxide and 21 percent oxygen, a mixture of 5 percent carbon dioxide and 21 percent oxygen, or normal air. The number of yawns and the duration of a yawn when it happened did not differ between any of these groups. Neither carbon dioxide gas nor 100 percent oxygen caused the students to yawn more. These gases also did not change the duration of yawns when they occurred.

Yawning and stretching seem to be related. A yawn is really a large stretch of the facial muscles. If you try to stop a yawn by clenching your jaws shut, you prevent the stretch, but also have a very unsatisfying yawn. Think about when you yawn most often; perhaps it is when you wake up and stretch your arms in the morning. Other data suggest that yawning may be nothing more than a mechanism to cool the brain.

Regardless of the reason for yawning, it is apparent that many people are susceptible to yawns if they see other people yawning, read about yawning, hear people yawning, or even think about yawning. In fact, at least half the population can "catch" a yawn from another person. Contagious yawning may be an evolutionary leftover from a time when nonverbal communication (the yawn) coordinated the behavior of a group of animals. Also, people who are aware of their own mental state and can see things from another person's perspective are more susceptible to contagious yawning. The ability to understand social cues also seems to be important. People with autism and schizophrenia have trouble understanding the point of view of other people, and they are much less likely to catch a yawn from someone else.

References

Chudler, E.H., Contagious yawning, https://faculty.washington.edu/chudler/yawnc.html, November 5, 2003, accessed February 3, 2016.

Gallup, A.C., and Gallup, G.G., Yawning as a brain cooling mechanism: nasal breathing and forehead cooling diminish the incidence of contagious yawning, *Evolutionary Psychology*, 5:92–101, 2007.

Platek S.M., Yawn, yawn, yawn, yawn; yawn, yawn, yawn! The social, evolutionary and neuroscientific facets of contagious yawning. *Frontiers in Neurology and Neuroscience*, 28:107–12, 2010.

Provine, R.R., Tate, B.C., and Geldmacher, L.L., Yawning: no effect of 3-5% CO_2, 100% O_2, and exercise, *Behavioral and Neural Biology*, 48:382–93, 1987.

? ? ? ? ?

Q

Can smiling make me happier?

A

If you are feeling sad and you would like to feel happy, fake it until you make it. Forcing yourself to smile can actually make you feel a little bit better. This in turn will make you smile, which will make you feel a little happier, and so on in a positive feedback cycle. The converse is also true: if you frown you'll feel a little less happy. Obviously this effect is limited; returning the smile of a friendly stranger does not send you into spasms of uncontrollable euphoria (usually) and you're probably not going to smile your way out of deeply sad situations.

Why does smiling give us an emotional boost? Because our brains are not as independent of our bodies as we tend to think. Our bodies are not merely fleshy vehicles that allow our brains to interact with the world. Instead, our bodies help our brains know what to think by setting the context in which information will be received. This is called *embodied cognition*. Psychologists have played with this in a variety of different ways. For example, if your body positioning (including your face) is positive, then you will be more inclined to like things and bad news will be less distressing. The opposite is also true. If you get Botox injections to paralyze your facial muscles (and minimize your wrinkles), you are less likely to feel anything at all. If you adopt power poses, you will act with greater confidence.

Your brain evolved with your body to form a unit. It's not surprising that they work together.

References

Davis, J.I., Senghas, A., Brandt, F., and Ochsner, K.N., The effects of BOTOX injections on emotional experience, *Emotion*, 10:433–40, 2010.

Niedenthal, P.M., Embodying emotion, Science 316:1002–5, 2007.

? ? ? ? ,

Q

Is there a "God spot" in the brain?

A

Religion is ubiquitous in human culture, but as far as we can tell, humans are the only animals who exhibit this behavior. This is important because it means that the capacity to have spiritual experiences is one of the things that makes humans unique. Scientifically speaking, if you are going to believe in anything, you have to do it with your brain. This naturally leads to the question: which part of the human brain believes in God? Is it possible to isolate one part of the brain that is specialized for supernatural phenomena?

The primary support for this idea comes from a type of temporal lobe epilepsy in which people experience hyperreligiosity. Hyperreligiosity is a form of religious mania characterized by one or all of the following: a sense of spiritual presence, extreme emotion (positive or negative), a feeling of transcendence, visual or auditory hallucinations, and the belief that you have been chosen to complete a specific task. Hyperreligiosity is not specific to any one faith and can manifest within any religious framework. It is also associated with some forms of mental illness, particularly psychosis. The question is: is hyperreligiosity just an extreme form of normal religious behavior? If so, perhaps the God spot is in the temporal lobes.

It is an interesting idea, but research suggests that religious experiences in normal healthy subjects are distributed across multiple brain areas, as would be expected for any higher-level cognitive function. In other words, there is no God spot.

Of course, this does not provide any insight into the existence or nature of God, whose existence is presumably independent of our belief or lack thereof.

Reference

Kapogiannis, D., Deshpande, G., Krueger, F., Thornburg, M.P., and Grafman, J.H., Brain networks shaping religious belief, *Brain Connections*, 4:70-79, 2014.

Is hypnosis real?

A

Hypnosis is strange, fascinating, mysterious, and makes a great plot device. As a result, most people are familiar with hypnosis on some level. Most people are also mostly misinformed about what it is. For example, it is not a sleep state, it doesn't necessarily involve amnesia, and you can't be hypnotized against your will.

Hypnosis is defined as an induced state of relaxation, attention, and suggestibility. A hypnotic state is often referred to as a *trance*. While in this trance, a hypnotized person will respond uncritically to suggestions and may experience hallucinations or illusory events, specific sensations, or the feeling that his body is out of his own control. These suggestions can persist after the trance has been terminated in an effect called posthypnotic suggestion. Some people are more hypnotizable than others for reasons that are not totally clear. Likewise, some people find the idea of hypnosis exciting, whereas others find it disturbing.

Hypnosis has two main uses: therapy and entertainment. The entertainment value comes from watching people do silly things on stage. The therapeutic benefits are related to behavioral modification (e.g., smoking cessation) or modulation of pain. Some people have had surgery using hypnosis as their only form of anesthesia. Hypnosis has also been used in psychotherapy. According to some theories, a person in this relaxed and uninhibited state is able to release painful, repressed memories. Unfortunately, the defining characteristic of hypnosis is suggestibility, and many people undergoing hypnopsychotherapy have ended up with implanted memories of childhood abuse. This underscores the potential power of hypnosis and the care that should be taken when it is employed as a therapeutic tool.

Although hypnosis has been known since the eighteenth century, the brain activity underlying hypnotic states is still not well understood. Part of the reason for this is because hypnosis has traditionally been the domain of psychology and has received little attention from neuroscientists. However, this is changing. The defining characteristics of a hypnotic trance make the frontal cortex a prime candidate for a neural substrate. Indeed, brain imaging has implicated a decoupling between the anterior cingulate cortex and the lateral frontal cortex, areas involved in conflict monitoring and cognitive control, respectively, in hypnosis. At this time, there is still more unknown than known, and much more research remains to be done. In the meantime, if it appeals to you, you can try it out and see if it helps you quit your bad habits.

References

Kihlstrom, J.F., Neuro-hypnotism: prospects for hypnosis and neuroscience, *Cortex*, 49:365–74, 2013.

Tobias, E., Jamieson, G., and Gruzelier, J., Hypnosis decouples cognitive control from conflict monitoring processes of the frontal lobe, *Neuroimage*, 27:969–78, 2005.

Is there a love hormone in the brain?

A

Prairie voles are famous (among neuroscientists) for being monogamous. These otherwise unremarkable rodents are unique among voles for their lifelong devotion to their mates. This is particularly interesting to humans, because we are also a monogamous species, which is rare among mammals.

What's behind that instinct to find a nice vole and set up a nest? Oxytocin—a chemical found in the brain and in a few other places in the body. In prairie voles injections of oxytocin or oxytocin blockers can create or prevent bonding, respectively. Vasopressin, a closely related hormone, is also important (especially in males). Certainly there are other factors, because pair-bonding is a complex social behavior. Oxytocin, however, has received most of the press. It has been widely popularized as the "cuddle hormone," but this is an oversimplification. Oxytocin is also known to facilitate parental bonding and behavior such as grooming and responding to babies' cries. In women, it is critically involved in childbirth and the perinatal period and causes uterine contractions as well as milk letdown during feeding. Many people are familiar with the synthetic version of oxytocin, Pitocin, which is used to induce labor and prevent postpartum hemorrhage.

Oxytocin has been implicated in a wide variety of human social interactions. For this reason it has been proposed as a treatment for autism spectrum disorder, a condition characterized by difficulty with social interaction and communication. Initial experiments have reported positive results: a single dose of nasally administered oxytocin temporarily enhanced social behaviors. Unfortunately, this improvement has failed to translate into a long-term effect. However, autism has many different subtypes, and it may be that only a small number of these types are treatable with oxytocin therapy.

In addition to the role that oxytocin plays in the nervous system, it is also involved with cardiovascular function, kidney function, and wound healing. Furthermore, it may provide a linkage between these systems and social behaviors. For example, social interaction may help you heal faster. All of this means that physiologically speaking, this is a really important hormone. It probably has functions that we don't even know about yet.

References

Shen, H., Neuroscience: the hard science of oxytocin, *Nature*, 522:410-12, 2015.

Young, L.J., The neural basis of pair bonding in a monogamous species: a model for understanding the biological basis of human behavior. In: National Research Council (U.S.) Panel for the Workshop on the Biodemography of Fertility and Family Behavior; Wachter KW, Bulatao RA, editors. *Offspring: Human Fertility Behavior in Biodemographic Perspective* (Washington, DC: National Academies Press, 2003).

How is brain research used in the courtroom?

A

Could a brain scan save a convicted murderer from the death penalty? Could testimony from a neuroscientist reduce the amount of time a criminal spends in jail? Could a defendant's claim of a head injury sway a jury's verdict? The answer to all three of these questions is "yes."

Brain research is making its way into the courtroom with increasing frequency. Dr. Nita Farahany, a law professor at Duke University, found 1,585 judicial opinions issued between 2005 and 2012 had used neurobiological evidence during legal proceedings. In 2005, about 100 of these opinions were issued, but in 2012, more than 200 of these opinions were issued. Neuroscientific evidence was used as part of criminal defenses in 60 percent of the 1,585 cases for serious felonies such as murder, drugs, assault, and robbery. Dr. Farahany estimates that neurobiological evidence is used in at least 5–6 percent of all murder trials in the United States.

Brain scans (magnetic resonance imaging, MRI; computerize axial tomography, CAT; positron emission tomography, PET) and brain recordings (electroencephalography, EEG) have all been introduced during court proceedings. Images of a defendant's brain are often used to show that the accused suffered a head injury or brain damage. Lawyers may also use the results of cognitive tests and expert testimony to paint a picture of the defendant.

Neuroscientific evidence is used to establish whether a person is mentally competent to go on trial. If a defendant does not have the ability to work rationally with a lawyer, then that person cannot go on trial. Defendants must be mentally competent to confess to a crime, and they cannot plead guilty unless they are mentally competent. The "brain scan defense" does not always work in favor of the defendant, but it is a tactic the defense lawyers are using more frequently to exonerate their clients or get reduced sentences.

Brain research has not provided much help to lawyers, judges, or defendants in determining whether someone is guilty or innocent of a crime. The polygraph or lie detector machine is a device that measures blood pressure, sweat, respiration, and heart rate and tries to correlate these physiological measurements to lying. Not only has the polygraph proved to be unreliable, but its use in the courtroom is either banned or severely restricted. Brain scanning methods are also not reliable in detecting lies, and it may be possible for people to suppress memories to fool such tests.

The U.S. Supreme Court has even considered neuroscience in some of its opinions. In deciding to ban capital punishment and life in prison without the possibility of parole for juveniles, the justices reflected on evidence showing that young brains are not developed like adult brains. Regardless of whether you are a Supreme Court justice or a member of jury, it is increasingly important that you know about brain research so you can properly weigh the evidence before rendering a decision.

References

Farahany, N.A., Neuroscience and behavioral genetics in US criminal law: an empirical analysis, *Journal of Law and Biosciences*, 1–25, 2016, doi:10.1093/jilb/lsv09.

Hu, X., Bergström, Z.M., Bodenhausen, G.V., and Rosenfeld, J.P., Suppressing unwanted autobiographical memories reduces their automatic influences: evidence from electrophysiology and an implicit autobiographical memory test, *Psychological Science*, 26:1098–106, 2015.

What is the relationship between the mind and the brain?

A

Clearly there is a relationship between the mind and the brain. We know this because if you destroy the brain, you destroy the mind. The body can be otherwise alive and healthy, but if the brain is dead, the person is dead. This is why brain death is synonymous with death, and why the organs of a brain-dead person can be harvested and used by others. This is not true for any other organ in the body.

Although we can confirm a robust relationship between the brain and the mind, we don't know much more about the nature of that relationship. Part of the reason for this is that the mind is difficult to define. If you can't even formulate a question, it is extremely hard to come up with an answer. How do you measure the mind? Most of us define consciousness the same way Supreme Court Justice Potter Stewart defined obscenity: we know it when we see it.

Of course, some scientists have tried to address the question of consciousness. As far as we can tell, the mind is an emergent property of the brain. This means that the individual cells are not conscious, but the interactions between them gives rise to consciousness. There is no "consciousness area" of the brain. Rather, different areas of the brain interact to produce self-awareness. This is why the mind is somewhat tolerant to brain damage. A fully functional brain is not necessary to have a mind. The important, and not at all hypothetical question is, how much (and which parts) of your brain do you need to be conscious? Sometimes we are not sure if a person is brain-dead, and knowing is a matter of life and death.

Is it possible that other animals have minds? It is not improbable. Is it possible that other, highly interconnected systems will achieve consciousness? How will we know? These are questions not only for scientists and physicians but for philosophers, ethicists, theologians, and legal scholars.

Does staring at inkblots really tell you anything meaningful?

A

It's a bird, it's a plane, it's . . . just a bunch of ink splotches. Actually, this is the Rorschach test, a psychological test created in 1921 by Swiss psychiatrist Hermann Rorschach (1884–1922). The Rorschach test consists of 10 cards spotted with symmetrical inkblots: five cards have black and gray spots; two cards have black, gray, and red spots; and three cards have pastel multicolored spots. During a Rorschach test, examiners ask people what they see on the cards. Only basic instructions are provided to avoid influencing a person's response. Everything a person says about a particular image, such as what the image on the card looks like, where the image comes from, and why they see a particular image, is recorded. Based on the results of the test, psychologists analyze a person's personality and thought processes, make assumptions about a person's unconscious desires and fears, and diagnose behavioral problems.

Although the Rorschach test is used by many therapists, its usefulness has been questioned based on the test's reliability and validity. Critics claim the test is not reliable because different people who score a particular test will arrive at different conclusions. Other detractors point out that the test is not valid because it does not predict many abnormal behaviors. This is especially true when the test is given to children and people in some ethnic minority groups. Also troubling is the lack of a clear normal Rorschach test result. The lack of a normal test result makes it difficult to know when a result is unusual or of concern. Although John E. Exner (1928–2006) updated the Rorschach scoring system in the 1960s, discrepancies between the results of the revised test and other psychological tests suggest that problems with the Rorschach test remain.

The Rorschach test is used around the world and depicted in movies and TV shows as a reliable way to uncover psychological problems. Despite this use, the Rorschach test remains controversial and the significance of its results are still open to interpretation.

References

Wood, J.M., Nezworski, M.T., Lilienfeld, S.O., and Garb, H.N., *What's Wrong with the Rorschach? Science Confronts the Controversial Inkblot Test* (San Francisco: John Wiley & Sons, 2003).

What happens to the brain during a near-death experience?

The term *near-death experience* (NDE) is very self-descriptive—it refers to something that some people experience as the result of a life-threatening condition during which they lose consciousness. About 15–20 percent of critical patients report an NDE, which is a sizable minority. NDEs cross gender, age, socioeconomic, educational, and religious bounds—they can happen to anybody, but they don't happen to everybody (at least not to everybody who recovers). The features of NDEs are similar regardless whom they happen to and include some combination of perceiving a tunnel, seeing a bright light, meeting deceased relatives, a feeling that one is dead, a life review, and an out-of-body experience. NDEs are usually (but not always) pleasant.

NDEs are intrinsically interesting because most of us would really like to know what will happen to us when we die. It is appealing to think that people who have briefly died can come back and tell us what it was like, especially when what they report is something nice. NDEs are interesting to scientists because if the central tenet of neuroscience is true, that we experience everything with our brains, then the NDE represents what happens to your brain when you are close to dying. These two perspectives create a fair bit of tension between people who believe NDEs represent a peek into the afterlife and those who believe they are evidence of nothing more than the brain in an interesting condition.

Who is right? It's basically impossible to know. First, the timing can't be nailed down. Everyone who has had an

NDE has lived to tell about it. Did the NDE happen as they were approaching death, during their brief "death," as they were being revived, or does it span all of these conditions? No one knows, and we're unlikely to find out because when someone is about to die, everyone is busy trying to stop that from happening. Usually, this sort of thing happens suddenly and unexpectedly, and no one is prepared to do a brain scan. Perhaps more important, the two hypotheses about the true meaning of NDEs are not mutually exclusive. If you prove that there is a physical basis in the brain for the NDE, that does not mean that it has no transcendent component, and vice versa.

All that being said, there are theories to explain what might be happening in the brain during an NDE. These include some combination of a loss of blood flow to the retina (tunnel vision), temporal lobe seizures induced by oxygen deprivation (meeting deceased relatives, life review), neurotransmitter imbalances, fear-elicited opioid release (feelings of euphoria), multisensory breakdown (out-of-body experiences), and previously formed expectations. None of these theories has been proven conclusively, and even taken together they do not thoroughly explain all the features of NDEs. However, they do provide a plausible neurophysiological basis for the NDE.

References

Facco, E., and Agrillo, C., Near-death experiences between science and prejudice, *Frontiers in Human Neuroscience*, 10:209, 2012, doi:10.3389/fnhum.2012.00209.

Mobbs, D., and Watt, C., There is nothing paranormal about near-death experiences: how neuroscience can explain seeing bright lights, meeting the dead, or being convinced you are one of them, *Trends in Cognitive Sciences*, 15:447–49, 2011.

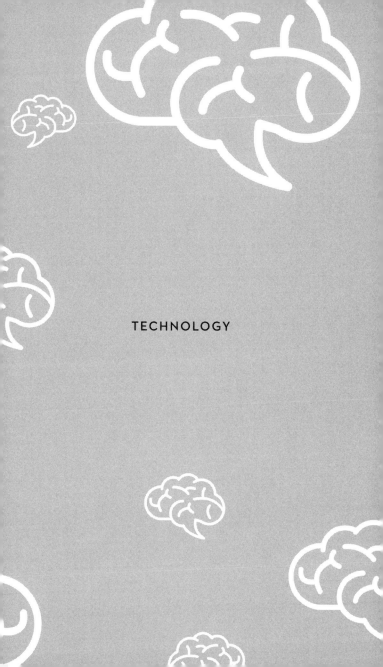

TECHNOLOGY

Can a computer be used to control my brain?

A

Sort of. Brain stimulation can be used to activate (or inactivate) certain parts of the brain. In modern times this almost always involves some sort of computer. This is exactly how a deep brain stimulator works to control the symptoms of Parkinson's disease, for example. Trials are currently under way to test the same technology as a therapy for depression and obsessive compulsive disorder. These interventions require implanted electrodes, but the brain can also be stimulated externally using electrical, magnetic, or ultrasonic stimuli. So, yes, a computer can be used to control movement, mood, and behavior. Whether that is a good thing or a bad thing depends on your perspective.

Before you panic, let's get some things straight. First, at least at present, your brain can't be controlled without your permission. You need to either have electrodes implanted or you need to put your head next to a stimulating device, and (legally) no one can force you to do that. It isn't possi-

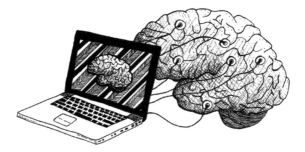

ble to control your brain through radio waves or a wireless network without your knowledge. So if you don't want to have your brain controlled, there is an easy way to avoid it. Second, controlling one part of the brain is not the same as controlling the entire brain, and brain control is not the same as mind control. Concerns about brain control may be legitimate, but we are not yet in the realm of a sci-fi/horror film, so you can take off the aluminum foil hat.

?   ?   ?   ?   ?

Q
—

Can scientists read a person's mind?

A
—

This is a tricky question because it gets a little philosophical. With the right equipment, scientists can decode some of what is happening in a person's brain. But what scientists can deduce is a long way from describing the complicated inner state that most of us would consider the "mind," For example, electrodes implanted in, on, or over the motor cortex can detect the neural signal to move some part of the body.

The extent of the information about when, how, and which part of the body is involved depends largely on where the electrodes are located: electrodes stuck into the brain yield much more information than electrodes stuck to the scalp. This signal can actually be detected before the onset of movement, and for that reason scientists sometimes say that they are decoding the intention to move. The signal is preserved even when someone has lost the ability to execute a movement; for example, when a person has a spinal cord injury or stroke. This means that it is possible to capture this movement intention and use it to control an external device, such as a computer or a prosthetic limb.

Is that mind reading? Maybe a little bit. This is far from the only piece of information scientists can read from your brain. By recording from different parts of your brain, scientists can tell if you have recognized something you were looking for or if you have made an error in the task you were trying to complete. Again, this isn't quite the image that comes to mind when you think about mind reading. Can scientists decode your stream of consciousness? No. Will it ever be possible? Maybe.

The mind is a sort of a gestalt property of the brain. It is difficult to define and difficult to study. As science and technology advance, it will certainly be possible to decode more of the brain's functions. That may translate into exciting opportunities, but it also implies an inherent loss of privacy. That is something worth worrying about. Fortunately, unlike mind reading, these functions can be defined, and thus anticipated and discussed openly and in advance.

? ? **?** ? ?

Q

Is it possible to see what's happening inside a living brain?

A

First of all, it is entirely possible to open up the skull of a living organism and physically look at the living brain. Neurosurgeons do this all the time. However, that provides a very limited amount of information. We really want to know what's happening in the brain. Therefore, we should be more precise with our language and ask, is it possible to visualize brain activity? The answer is: yes, sort of.

A variety of different techniques can map which parts of the brain are active under different conditions. Collectively these techniques are referred to as neuroimaging. Although all of these techniques measure ongoing brain activity, there are some significant differences between them. The first (and perhaps the most important) is the type of activity the different methods measure. Some techniques like electroencephalography (EEG) and magnetoencephalography (MEG) measure a signal generated by the neurons. These are direct measurements of what the brain is doing. Other techniques such as functional magnetic resonance imaging (fMRI), positron emission tomography (PET), or single-photon emission computed tomography (SPECT) measure things that are related to neural activity (like blood flow or metabolism), but not the neural activity itself. As such, these methods are one step removed from what the brain is actually doing. Another important difference in these techniques is how fast they can measure changes. EEG and MEG measure electricity, which moves almost as fast as the speed of light, and can keep up with neural activity in real time. Other techniques are much slower, sometimes taking data samples only every few seconds, which is slow compared to the speed of the brain. Finally, there is a difference in what parts of the brain the different techniques can measure. EEG and MEG can only measure what is happening on the brain's surface, which can be quite limiting. PET, fMRI, and SPECT, on the other hand, can provide information about the whole brain.

Clinically, EEG is used for a broad range of monitoring and diagnostics. It is well suited for epilepsy monitoring, for example, because it can record rapid neural activity. PET is primarily used for cancer detection. SPECT imaging is also used in oncology and to image some diseases, like Parkinson's, that affect deep parts of the brain. fMRI is used for mapping brain function prior to surgery.

The newest trend is to use several of these techniques simultaneously (which can be technologically challenging) to get as much information as possible.

Are brain transplants possible?

A

No, but head transplants are theoretically possible. Actually, it is unclear whether they should be called head transplants or whole body transplants. Either way, what you end up with is one person's head attached to another person's body. These are only theoretically possible in humans. However, in a set of strange and unsettling experiments, the feasibility of head transplantation has been demonstrated in animals.

The first experiments in head transplants involved attaching the head of a decapitated dog to the neck of an intact dog, using the second dog's blood supply to support the severed head. These types of experiments have been performed a number of times by different research groups, and they have been surprisingly successful (both dogs lived for a few days before succumbing to immune reactions) especially given that they were performed in the early to mid-1900s. In the early 1970s, Robert White, a neurosurgeon at Case Western Reserve, performed a successful whole head transplant on a monkey. Following an 18-hour surgery, the animal lived for several days and was able to see, hear, taste, and smell and even tried to bite lab assistants. White performed a second head transplant on a monkey in 2001 and expressed hope that the surgery could someday be performed on humans whose own bodies were succumbing to terminal illness.

Of course, in such a restored life, a person would be paralyzed from the neck down and on perpetual life support. As there is currently no way to connect the brain of the transplanted head to the spinal cord of the donor body, the body is really only acting as nutrient support for the transplanted head. Even for those seeking immortality, this may not be the kind of endless life they have in mind.

The idea of a head transplant has been met with almost universal horror and disgust in both public and scientific circles. In spite of these sentiments, the first human head transplant has been scheduled for 2017. In a very controversial and high-profile case, Italian neurosurgeon Sergio Canavero has announced plans to perform a head transplant in China on a Russian man suffering from a muscle wasting disease. Canavero claims that he will be able to link the spinal cord to the head, giving the patient control over the donor body. It remains to be seen if the surgery will actually happen or if it will be successful. If this surgery is performed, will it be a scientific miracle or total abomination?

Reference

Canavero, S., HEAVEN: The Head Anastomosis Venture Project outline for the first human head transplantation with spinal linkage (GEMINI), *Surgical Neurology International* 4(S1):S335–42, 2013.

Can sight be restored to someone who is blind?

A

To some degree, yes. Several types of visual implants have been designed to interface with the nervous system at different points along the visual pathway. The important thing to remember is that you see with your eyes, but you perceive with your brain.

Electrical stimulation of neurons at any point in the visual system will cause a visual perception, but the features of that perception might be different depending on where you stimulate. Light is transformed into a neural signal when it falls on the retina, the layer of tissue (actually a part of the brain) on the inside surface of the back of the eyeball. The visual signal leaves the eye through the optic nerve and makes a stop in the thalamus before it ends up in the visual cortex.

The retina is the most frequently targeted site for a visual prosthesis for a couple of reasons. First, it is early in the visual pathway, which means you can (theoretically) take advantage of most of the brain's native signal processing. Second, it doesn't require brain surgery, a serious risk for an elective procedure. The idea behind a retinal implant is to replace the light-sensing cells (photoreceptors) in the retina by stimulating the downstream neurons directly. An externally worn camera acquires visual information which is then processed by an onboard computer and sent to an array of electrodes in the retina. Although the number of electrodes is much smaller than the number of photoreceptors in a healthy retina, people with these implants can use a retinal implant for navigation, object discrimination, and even reading large print. This is because the brain is phenomenal at using any kind of information you give it. The FDA approved the first retinal implant, the Argus II, for use in the United States in 2013.

Of course, a retinal implant requires that you have a retina. Two of the leading causes of blindness in the United States are macular degeneration and retinitis pigmentosa. In both conditions, the photoreceptor cells die while sparing the rest of the retina. So a retinal implant would be appropriate. But if you are blind because you have lost your eyes, retinal implants will leave you in the dark. You can still stimulate elsewhere in the visual pathway, however. In a series of high-profile cases in the early 2000s, a neuroscientist named William Dobelle contracted a neurosurgeon to implant cortical visual prostheses in paying customers. Several of these people were able to regain useful vision. One patient even drove a car (slowly, in a parking lot for a group of reporters). Unfortunately, the project was fraught with ethical issues. Dobelle himself was quite ill and died suddenly without designating anyone to care for his patients or take over his work after his death. All of the implants stopped working after a relatively short period of time. Worse still, the implants caused such serious health problems that all of Dobelle's customers were forced to pay to have the devices removed. This might put you off cortical implants, but remember, they did work. Other groups are continuing to pursue research on cortical visual implants in a much more responsible manner. In addition, an optical nerve prosthesis has been developed and a thalamic prosthesis has been proposed.

What is a cochlear implant?

A cochlear implant is a device that uses electrical stimulation to compensate for profound hearing loss. It is only effective in people who are deaf as a consequence of damage to a set of specialized sound-transducing cells (hair cells) in a part of the inner ear called the cochlea. Even so, the device is quite useful because most cases of hearing loss are related to hair cell death or congenital malformation.

Hair cells can be damaged by loud noise, illness, and even some medications. Unfortunately, hair cells never regenerate: once they are gone, they're gone (something to think about next time you reach for your headphones). Cochlear implants are designed to replace the hair cells by detecting sound and turning it into electrical impulses the nervous system can interpret. This turns out to be a manageable problem because the cochlea is laid out such that the hair cells sensitive to high frequencies are at one end and those hair cells sensitive to low frequencies are at the other end. All the intermediate frequencies fall in between in a neat line. This organization is called a *tonotopic map*. Knowing the mapping means that scientists know where to stimulate to elicit the perception of a particular sound.

During a cochlear implant surgery, a long thin array of electrodes is placed in the cochlea. These electrodes stimulate the cochlear nerve based on a signal that is picked up by an externally worn microphone. Surprisingly, this works pretty well. Although the number of implanted electrodes is very small compared to the number of hair cells in a healthy cochlea, people who use cochlear implants are able to understand speech, even in the absence of visual cues (on the telephone, for example). This remarkable fact underscores the brain's startling capacity to make use of any information it can get.

Cochlear implants are useful for people who have acquired hearing loss as well as for those who were born deaf. People who are born deaf should get the implant very early in life—ideally by two years of age. This is because the brain goes through a so-called critical period for auditory information early in development. If the brain doesn't receive any auditory inputs during this period, the window of opportunity for learning to process that information closes. This brings us to something controversial about cochlear implants: not everyone thinks they're a great idea. Notably, some members of the deaf community take exception to the idea that deafness is a condition that needs to be fixed, and they don't believe it is appropriate to subject children to what is essentially an elective surgery. This is exacerbated by the fact that most deaf children are born to hearing parents, meaning the decision to have the surgery will be made by someone who doesn't have a firsthand understanding of what it is like to be deaf and who isn't part of the deaf community. Unfortunately, it isn't possible to wait until the child is old enough to decide for him- or herself because at that point it will be too late. This puts parents in a difficult position. Ultimately they must decide what is best for their child.

? ? ? ? ?

Q

What is a brain-computer interface?

A

Brain-computer interfaces (BCIs) are devices that directly connect some part of the nervous system with the outside world. They are incredibly cool and deeply unnatural.

BCIs don't have to involve either the brain or a computer, so it's a bit of a misnomer. You might hear these devices referred to as brain-machine interfaces or brain-controlled interfaces. This is primarily because scientists don't like to use terms that other people have coined, and not because they are inherently more useful. "Direct neural interface" is a more accurate but less common name; a "neuroprosthetic" is essentially the same thing with a more medical angle.

Terminology aside, what do these devices actually do? Remember that your nervous system is usually connected to the world through the rest of your body. Your muscles contract, either volitionally or automatically, in response to the commands your nervous system generates, and in this way you are able to exert forces on the environment. You can move around, pick things up, point, and flex. In a similar way your sensory organs—your eyes, ears, nose, skin, tongue—detect important physical properties about the outside world (and inside your body) and convert them into neural signals that are processed by the nervous system. Your body is really good at doing this, but that doesn't mean it's the only show in town. We have other ways of exerting forces on the environment (through robots, for example) and other ways of sensing what's happening in the world (such as with a thermometer). If we can figure out how to get information in and out of the nervous system, then we don't need our bodies anymore.

You might wonder why that would be of real interest to anyone. If "because we can!" isn't a sufficiently compelling reason for you, then consider the medical implications. As great as our bodies are when they are working properly, they don't always work properly. They can be damaged by injury or disease, and this can restrict our options for interacting with the world. An example that many people are familiar with is a spinal cord injury that can result in paralysis (inability to move) and loss of sensation. There are many other examples, including stroke, neurodegenerative diseases, traumatic brain injury, and amputation. The special senses, such as vision and hearing, can also be disrupted by everything from high fever to cancer to loud music.

Most BCIs help assist or rehabilitate people in some way: they restore vision or hearing, allow locked-in people to communicate with a computer, provide controls for sophisticated robotic arms, or restore balance to people with vertigo. Some of these devices are noninvasive, meaning they attach to the outside of the body and are easily removed. Many BCIs rely on some sort of implant and require surgery to use. One thing they all have in common is that they are not as good as the real thing: they don't perform nearly as well as the body parts they are intended to replace. However, they can do some amazing things, and they are getting better all the time. It's not inconceivable that people will want to use them for enhancement. Whatever you want to call them, BCIs will likely be an important part of our future.

? ? ? ? ?

What is brain stimulation?

A

The idea behind brain stimulation is to force, at the moment of your choosing, some set of neurons to become active by using an external signal. The general motivation for doing this is to create a response in the person whose brain you are stimulating. In many cases, the response itself is desirable in some way. For example, stimulation might cause the perception of light, sound, or touch or might stop a tremor or cause the person to feel less depressed. In other cases the response helps you understand what the stimulated part of the brain does in real life. For example, if a stimulus causes your leg to twitch, then it is (mostly) justifiable to believe that the area stimulated is involved in leg movement.

That pretty much sums up the "why" of brain stimulation, which leaves us with the "how" question. Traditionally, the answer has been electricity. When it comes to brain stimulation, electricity has some useful properties. First, we've known about electricity for a long time. We know how to make it, measure it, and get it where we want it to go. That's tremendously helpful. But more important, the brain already uses electricity as its native language for internal communication. We are a long way from understanding or speaking that language fluently, but we can communicate a few rough points. The clarity of those messages depends on where the electrodes are (outside the head, on the surface of the brain, or deep in the brain) and how you stimulate. There are also drawbacks ranging from burns to brain damage. So electrical stimulation of the brain isn't a perfect technique, and it is not something that should be tried at home.

Of course, there are other ways to stimulate the brain. The most basic form of brain stimulation is mechanical. Mechanical stimulation of the brain seems to come up a lot in popular imagination, but is rarely used (if ever) in modern clinical or research practice. Magnetic stimulation, which is a close cousin of electrical stimulation, is frequently used as a noninvasive form of brain stimulation. More recently, ultrasound has been used for noninvasive brain stimulation. Note that in this context *noninvasive* means that it doesn't require any surgery, not that it is inherently safe. You can also stimulate the brain with light, although most neurons don't naturally respond to light (it's pretty dark inside your head), so this takes some genetic fiddling. For this reason it is referred to as *optogenetics*. The ethics of genetic manipulation are complicated, which is why, for the moment, this technique is not used in humans.

In general, brain stimulation is a useful tool, but a complicated beast. There is still a lot we don't know about how it works and how it will impact the entire nervous system in the long term.

MEDICINE

What is schizophrenia?

Schizophrenia is potentially debilitating mental illness characterized by psychosis, or loss of contact with reality. It is commonly confused with, but not the same as dissociative identity disorder, in which people exhibit multiple distinct personalities. The indications of schizophrenia include positive symptoms such as hallucinations (usually auditory), delusions, and disorganized thinking as well as negative symptoms such as social withdrawal, self-neglect, loss of motivation, and flat affect. In this context *positive* and *negative* are not value statements. Rather, they indicate whether a behavior is adopted or lost. The types of hallucinations and delusions experienced by people with schizophrenia tend not to be pleasant and uplifting. On the contrary, they are usually malignant and terrifying.

Symptoms typically appear in early adulthood and may initially be difficult to distinguish from normal teenage behaviors. Antipsychotic medications can treat the symptoms of schizophrenia. But there is no cure, and most people who live with illness deal with it for the rest of their lives. Unfortunately, people who suffer from schizophrenia cannot appreciate that their hallucinations and delusions are not real. As a consequence, it can be difficult to convince them to seek the help they need.

The biological underpinnings of schizophrenia are still a bit of mystery, although the disorder seems to be related to the neurotransmitter dopamine. There is certainly a genetic component to schizophrenia, but environmental factors also play a role. Close relatives of people with schizophrenia are at increased risk for developing mental illness (including but not limited to schizophrenia). But, even identical twins of people with schizophrenia are not guaranteed to manifest symptoms themselves. Men are more likely to develop symptoms than are women, and the symptoms exhibited by men are more severe than those in women. The prognosis is usually less positive for men, too. People with schizophrenia are more likely to have experienced complications during the prenatal and perinatal period, and the disorder is more common in migrants and people who are born in and live in large cities. Drug use, especially cannabis and stimulants like cocaine and amphetamines, can also increase the risk of developing schizophrenia in people who are predisposed to the disorder. This body of evidence suggests that schizophrenia arises from a complex interaction between genes and environmental factors.

It is interesting to note that there are mild forms of the disorder, and otherwise healthy people can suffer psychotic episodes. This leads scientists to believe that schizophrenia is a disease in which a normal process becomes excessive. In other words, the difference between a normal state and a disordered state is a matter of degree.

Reference

Picchioni, M.M., and Murray, R.M., Schizophrenia, *British Medical Journal*, 335:91-95, 2007.

What is prion disease?

A

Prion diseases, also known as transmissible spongiform encephalopathies (TSEs), are a family of rare neurodegenerative disorders that are always progressive and always fatal. They can have long incubation periods (up to 40 years), but the decline is rapid after the first sign of symptoms which include changes in personality, cognitive impairment, and abnormal movements.

TSEs can be inherited or acquired and are communicable between humans and animals. The prion disease most people are familiar with is bovine spongiform encephalopathy, commonly known as mad cow disease. This disease can be transmitted to humans through consumption of beef contaminated with nervous tissue. (As a rule, it is never a good idea to eat nervous tissue.) Other TSEs include chronic wasting disease and scrapie in animals and Creutzfeldt-Jakob disease, fatal familial insomnia, and kuru in humans.

What makes prion diseases simultaneously interesting and horrifying is that the infectious agent is a protein (prion protein, PrP), rather than a living organism like a virus or a bacterium. This distinguishes TSEs from every other infectious disease known to humans. In normal, healthy people, PrP is found in the central nervous system as well as in the heart, lung, and lymphatic systems, but it is unclear what its natural function is. In prion disease, PrP is misfolded into a very stable but pathological conformation. What is remarkable is that the misfolded protein induces other healthy proteins to adopt the same pathological conformation—they are self-templating. The newly malformed proteins then convert other proteins, and so on, in an exponential manner. The mechanism for this contagious quality is not fully understood. The misfolded PrP aggregates and forms plaques that accumulate, first in lymphatic tissues and later in nervous tissues. These plaques are neurotoxic, leading to cell death and the development of holes in the cerebral cortex, giving it a spongy appearance, which is why these disorders are called spongiform encephalopathies.

Unfortunately, because prions are not alive, they cannot be killed, and because they are so stable, they are resistant to destruction by physical and chemical agents. There are currently no approved treatments for TSEs.

Prion disease is not the only neurodegenerative disease connected with protein misfolding. Alzheimer's disease, Parkinson's disease, and Huntington's disease are all similarly characterized by the misfolding and aggregation of proteins. Like prions, these proteins appear to be self-templating. It is unclear whether this type of conformation-changing cascade is always connected to disease or if these are cases of useful processes gone bad.

References

King, O.D., Gitler, A.D., and Shorter, J., The tip of the iceberg: RNA-binding proteins with prion-like domains in neurodegenerative disease, *Brain Research*, 1462:61–80, 2012.

Roettger, Y., Du, Y., Bacher, M., Zerr, I., Dodel, R., and Bach, J-P., Immunotherapy in prion disease, *Nature Reviews Neurology*, 9:98–105, 2013.

What is Hippocrates's "sacred disease?"

Answer: Ancient Greek philosopher Hippocrates spent considerable time contemplating the brain. He wrote about a condition where people would shiver, froth at the mouth, and stop speaking. Modern-day scholars believe he was describing epilepsy.

Epilepsy is not a disease so much as it is a symptom. A diagnosis of epilepsy simply means that you have recurring, unprovoked seizures but does not address the underlying cause of the seizures. A seizure is the physical manifestation of what is essentially an electrical storm in the brain. During a seizure, the normal balance of electrical activity that makes your brain function swings out of control becoming (temporarily) excessive and overly synchronous.

There are different types of seizures, and a multitude of different syndromes that can cause epilepsy. The symptoms of a seizure can be very dramatic, including convulsions, frothing at the mouth, and teeth clenching. However, the symptoms may be so slight as to be almost unnoticeable, appearing as periods of staring or "spacing out." Sometimes seizures are preceded by an aura, a set of warning signs that a seizure is imminent such as a strange taste, a sense of déjà-vu, or a fearful feeling.

There are two major categories of seizures: generalized and partial. Generalized seizures start at the same time across the cortex, whereas partial seizures start in one area and spread. Generalized seizures are more likely to have a genetic (heritable) origin. For example, some generalized seizure disorders are the result of mutations in ion channel proteins that are expressed throughout the brain. Partial seizures, on the other hand, are more likely to result from brain injury. A hard blow to the head, an infection, a tumor, or a period of oxygen deprivation could do it, although sometimes the source of the injury is never identified. There are also some genetic conditions that cause partial seizures.

Interestingly, in acquired epilepsy there is often a period of months or years between the event that causes brain injury and the onset of seizures. This suggests that the development of epilepsy is a gradual process, which presents the possibility that a timely intervention could prevent seizures from ever developing. Unfortunately, no preventive measures have been identified to date.

Several drugs can be used to control epilepsy, although many have adverse side effects. People with refractory partial epilepsy may be candidates for a brain surgery to reduce or eliminate their seizures. In some rare cases, an entire half of a child's brain may be removed or disabled to treat severe and otherwise uncontrollable epilepsy. Other management strategies include identifying and avoiding potential triggers like flashing lights, sleep deprivation, energy drinks, and in one unfortunate case, the sound of a particular TV personality's voice. Some forms of epilepsy, especially heritable childhood epilepsies, will spontaneously resolve.

It is also possible to have a seizure and never go on to develop epilepsy. This is especially common in children who experience febrile, or fever-related, seizures.

Reference

Chang, B.S., and Lowenstein, D.H., Epilepsy, *New England Journal of Medicine*, 349:1257-66, 2003.

Can marijuana be used to treat epilepsy?

A

Yes, marijuana can and has been used to treat epilepsy. The marijuana plant, known as *Cannabis*, contains a large number of chemical substances that bind to receptors in the brain. These chemicals are called cannabinoids or phytocannabinoids when they are derived from plants. When ingested or inhaled, these chemicals interact with the body's endocannabinoid system (the naming conventions here are not coincidental), which has a neuromodulatory function.

The two primary phytocannabinoids found in marijuana are tetrahydrocannabinol (THC) and cannabidiol (CBD). It has been known since the 1970s that both of these substances are anti-convulsive. If you have heard of THC, it is probably because of its more well-known psychotropic properties: THC is the part of marijuana that produces the "high." Although this can be recreationally attractive, it's really a drawback when it comes to medicinal use. CBD on the other hand, is not psychotropic, and evidence suggests that it is safe (nontoxic).

Inspired by these facts, Paige Figi took treatment of her daughter's severe and intractable epilepsy into her own hands. Charlotte Figi suffered from Dravet syndrome, which caused her, at the age of five, to have up to 50 generalized tonic-clonic seizures every day. Having that many seizures is catastrophic (that's actually the medical term for it), and it had a significant detrimental impact on Charlotte's development and quality of life. Paige lived in Colorado, where medical marijuana was legal, and she was able to find growers who had bred a strain of the plant with high concentrations of CBD and low concentrations of THC. After clearing her plan with Charlotte's doctors,

Paige gave gradually increasing doses of the preparation to Charlotte. The result was a 90 percent reduction in the number of seizures Charlotte experienced. The difference was absolutely staggering. Charlotte's story was featured in a 2013 documentary that inspired a surge of epilepsy-related immigration to Colorado. So far the evidence for the effectiveness of this treatment is anecdotal, but positive, and for many desperate parents, that is good enough.

The problem here is that self-medicating is a tricky issue. There are no established dosage guidelines, and it is not clear how marijuana might interact with other medications (of which epilepsy patients take quite a few). There is currently no federal or state oversight of marijuana growers and the chemical composition of the product is not regulated or guaranteed. Insurance companies do not cover medical marijuana, and the treatments can be quite expensive. Additionally, the strain of marijuana that was used to treat Charlotte is proprietary. Because marijuana cannot be legally transported across state lines, you have to live in Colorado to get it. A new drug called Epidiolex, which is made from pure plant-derived CBD, is now in clinical trials. However, pure CBD is not the same as a strain of marijuana that is high in CBD. There are still other substances in the whole plant, and these other chemicals may be critically important for its therapeutic effects. All of these factors make the decision to use marijuana difficult, especially when you are making the decision for a child. On the other hand, for many children, the alternative is literally catastrophic.

References

Epilepsy Foundation Colorado, FAQs, http://www.epilepsycolorado. org/index.php?s=10907, accessed January 5, 2016.

Maa, E., and Figi, P., The case for medical marijuana in epilepsy, *Epilepsia*, 55:783–86, 2014.

What is Alzheimer's disease?

A

Alzheimer's disease (AD) is a neurodegenerative disorder that attacks areas of the brain responsible for memory, language, emotion, and decision making. As the disease progress, people with AD develop dementia and a general loss of intellectual abilities. Eventually, day to day activities become more difficult, and a person with AD may need help eating, bathing, and getting dressed. Memory problems are common in people with AD and can progress to the state where a person will not recognize friends or relatives and may not be able to hold a conversation with others. Often people with AD can become withdrawn or depressed; they may also wander and get lost trying to find their way back home.

The brains of people with AD show several common features, including amyloid plaques, neurofibrillary tangles, and cell loss. Amyloid plaques consist mainly of beta amyloid proteins. When too much beta amyloid builds up, it forms plaques, a sticky build-up around neurons. Neurofibrillary tangles, found inside neurons, are abnormal clumps of a different protein called tau. The plaques and tangles interfere with the normal operations of neurons and eventually result in widespread damage to the brain, especially in areas such as the cerebral cortex and hippocampus. As a result, neurons fail to function normally and eventually die.

Although there is no cure, some drugs can be used to treat the symptoms of mild to moderate AD. Some of these drugs, such as donepezil, work to slow the breakdown of the neurotransmitter acetylcholine and can improve memory problems experienced by people with AD.

On November 5, 1994, former U.S. President Ronald Reagan released a letter to announce that he had Alzheimer's disease. In the short letter, Reagan thanked the people for the honor of serving as president and expressed hope that his disclosure would raise awareness about Alzheimer's disease.

He used the future tense in his letter when he wrote, "I have recently been told that I am one of the millions of Americans who will be afflicted with Alzheimer's Disease." Based on his speech pattern, some researchers suggest that he showed signs of the disease while he was still in office.

Reference

Berisha, V., Wang, S., LaCross, A., and Liss, J., Tracking discourse complexity preceding Alzheimer's disease diagnosis: a case study comparing the press conferences of Presidents Ronald Reagan and George Herbert Walker Bush, *Journal of Alzheimer's Disease*, 45:959–63, 2015.

What is encephalitis?

A

Encephalitis is a serious condition caused by an acute inflammation (swelling) of the brain. Inflammation can lead to cell loss, bleeding, brain damage, and death — all things that are best avoided. The infections that lead to encephalitis can be viral, bacterial, or parasitic. For example, encephalitis is how the rabies virus kills you. How can you avoid getting rabies? Don't pet or feed wild animals, stay away from stray animals, make sure your pets are vaccinated, and make sure you get prompt medical treatment if you are bitten or scratched by any animal that you are not confident has been vaccinated. Go to the doctor if you wake up in a room with a bat. Seriously – treat all bats like they have rabies.

Other diseases that can cause encephalitis include measles, mumps, polio, rubella, and varicella (chicken pox). You can avoid all of these by getting vaccinated. Unfortunately, there are other infections for which vaccinations do not exist, such as bacterial and parasitic infections. In these cases, good hygiene and pest control are your best defenses.

If one survives the acute phase of encephalitis, brain damage may still result in long-term consequences. Such effects are best illustrated by the interesting and terrifying case study of encephalitis lethargica. This disease was epidemic in Europe and North America in the 1920s; millions of people died, and many more were disabled. Following recovery from the acute phase of the disease, many adults suffered motor dysfunction, including symptoms similar to those observed in people with Parkinson's disease. These symptoms sometimes didn't show up until months or years after the initial infection and could be very severe. Some people became permanently catatonic—statue-like—and had to be institutionalized. In children, encephalitis lethargica often caused serious mental and behavioral problems. Some of these young survivors were transformed from completely normal children into dangerous psychopaths, challenging contemporary views of criminal culpability. Later, when these unfortunate survivors reached adulthood, postencephalitic Parkinsonism set in. Oliver Sacks's book *Awakenings* is about some of these patients. One of the most disturbing aspects of encephalitis lethargica is that the infectious agent is still unknown. The epidemic went away, but there is no reason it couldn't come back. A few cases still show up in hospitals today.

References

Dickman, M.S., von Economo encephalitis, *Archives of Neurology*, 58:1696–98, 2001.

Ruiz, V., A disease that makes criminals: encephalitis lethargica (EL) in children, mental deficiency, and the 1927 Mental Deficiency Act, *Endeavour*, 39:44–51, 2015.

What disorder affected actor Robin Williams?

Although Robin Williams was a prolific and active actor until his death in 2014, he was plagued by hallucinations, movement problems, anxiety, and depression. The autopsy examination of his brain revealed the cause: a neurological disorder called Lewy body dementia (LBD).

Although most people have not heard of LBD, this disease affects approximately 1.4 million people in the United States. LBD is difficult to diagnose because its symptoms are shared by several other diseases. For example, Williams was diagnosed initially with Parkinson's disease because he had trouble moving. Some people with LBD appear to have Alzheimer's disease because they get confused or forget things. In addition to having depression, visual hallucinations, and movement problems, people with LBD may have trouble thinking, sleeping, solving problems, and paying attention.

The only way to confirm the diagnosis of LBD is to examine the brain for small protein deposits called Lewy bodies. Blood tests and brain imaging can rule out other disorders that are causing the problems. LBD is thought to result from clumps of Lewy bodies that prevent the brain from properly sending signals in areas associated with movement and memory. Lewy bodies are also found in the brains of people who have Parkinson's disease and Alzheimer's disease. Although scientists do know not the cause of LBD and there is no cure, some medications can help reduce the cognitive, mood, and movement symptoms of the disease.

References

Lewy Body Dementia Association, http://www.lbda.org/, accessed January 7, 2016.

Walker, Z., Possin, K.L., Boeve, B.F., and Aarsland, D., Lewy body dementias, *Lancet*, 386:1683–97, 2015.

? **?** **?** **?** **?**

Q

What happened to Christopher Reeve, the actor who played Superman in the movies?

A

On May 27, 1995, Christopher Reeve was competing in an equestrian event when his horse stopped suddenly in front of a fence jump. Reeve was thrown head-first over the horse and landed on his head. Although Reeve was wearing a helmet and other protective gear, the force of the accident crushed his first and second cervical vertebrae (bones in his upper neck). The broken bones compressed the spinal cord that runs through the middle of the vertebrae.

The spinal cord carries messages from the brain down to the body to control movement and brings sensory signals about touch, temperature, pressure, and pain to the brain.

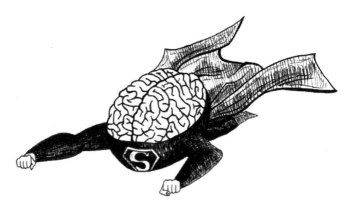

Because of the injury, Reeve was paralyzed from the shoulders down and unable to feel anything below the injury. He also needed a ventilator to help him breathe. Surgery was performed to repair the vertebrae and stabilize his skull.

Several years after the accident, Reeve began a rehabilitation program involving functional electrical stimulation (FES) and aqua therapy. During the program, he had electrodes placed on his legs to help him pedal a stationary bicycle. In addition to using FES, he worked out in a pool. After several years of therapy, Reeve showed modest improvement, including better neck function, movement of some finger joints, and some ability to detect touch on his skin.

In the early 2000s, Reeve had several infections and an ulcer. In 2004, his heart stopped and he went into a coma; he passed away on October 10, 2004, at the age of fifty-two.

Reference

Reeve, C., *Still Me* (New York: Random House, 1998).

? ? Q ? ? ?

What disorder does actor Michael J. Fox have?

A

Michael J. Fox has a movement disorder called Parkinson's disease. Parkinson's disease is a neurodegenerative disorder that occurs when neurons that produce the neurotransmitter dopamine die. Neurons that make dopamine are connected to neuronal circuits that help control movement. When enough of the dopamine-producing neurons die, electrical activity within these circuits becomes unbalanced and people have trouble moving.

The common symptoms of Parkinson's disease are slow movement; shaking or tremor of the hands, legs, and arms; stiff arms and legs; muscle cramping; and poor balance. Therefore, people with Parkinson's disease may have trouble walking, talking, writing, and doing everyday things around the house and at work. The cause of Parkinson's disease is not known, and although some cases may have a genetic component, the disorder does not seem to run in families. Although there are no cures for Parkinson's disease, medications that enhance brain levels of dopamine do provide some relief from the disorder's symptoms. Unfortunately, the benefits of dopamine medications decline as the disease marches on.

Michael J. Fox was diagnosed with Parkinson's disease in 1991. Fox told the public about his condition and underwent brain surgery. During surgery, Fox was awake while neurosurgeons inserted a small wire electrode into the region of the brain called the thalamus. The surgeons passed electrical current through this wire to permanently destroy neurons to rebalance the neural activity in the circuits responsible for movement. Other people with Parkinson's disease opt for deep brain stimulation (DBS), where pulses of electricity are sent through a surgically implanted electrode. The electrical pulses disrupt the neural signals causing the movement problems. These treatments do not cure Parkinson's disease, but they do often reduce the symptoms.

Because Parkinson's disease often progresses slowly, people affected by the condition can still have productive lives. Fox is active in his community and is involved with charitable activities, especially raising funds for new treatments and cures for Parkinson's disease.

References

Fox, M.J., *Always Looking Up. The Adventures of an Incurable Optimist* (New York: Hyperion, 2009).

Fox, M.J., *Lucky Man. A Memoir* (New York: Hyperion, 2002).

What neurological problem did many US Presidents share?

A

U.S. Presidents John Tyler, Andrew Johnson, Chester Arthur, Franklin D. Roosevelt, Dwight D. Eisenhower, Richard Nixon, and Gerald Ford had one. Warren G. Harding may have had one. John Quincy Adams and Millard Fillmore had two each. Woodrow Wilson likely had several. What did all of these statesmen have? They all had at least one stroke, also called a brain attack.

Although the brain makes up only 2 percent of the body's weight, it uses approximately 20 percent of the body's blood supply. The blood brings fuel, oxygen, vitamins, and minerals to support and nourish cells of the brain. A stroke happens when the blood supply to the brain is cut off. Without fuel to support cellular function, neurons start to die. When neurons start to die, function associated with those neurons is lost. For example, if a stroke occurs in the motor cortex, a person could lose the ability to move an arm or a leg; if a stroke occurs in an area of the brain responsible for language, a person could lose the ability to speak.

Blood flow to the brain may be compromised when a blood vessel in the brain or neck becomes blocked. The blockage could be caused by a blood clot or a narrowing of an artery. Strokes can also happen when a blood vessel breaks.

The best defense against stroke is prevention. High blood pressure, smoking tobacco, obesity, and lack of exercise all increase the risk of stroke, so lifestyle choices are important factors. If a stroke occurs, time is of the essence, and immediate medical attention is essential to reduce long-term damage and disability. For strokes caused by blocked blood vessel, a tissue plasminogen activator drug may be given to dissolve the clot and restore blood flow to the brain. A blood clot may also be cleared mechanically with a procedure called a thrombectomy. A mechanical thrombectomy involves threading of a small wire up to the location of the blood clot. The blood clot is then gripped by the wire and removed or vacuumed out so blood can flow through the vessel. Surgery can also be performed to stop the bleeding from a ruptured blood vessel.

Remember FAST to know the signs of someone having a stroke:

F: Face drooping: a person's face droops or is numb.

A: Arm weakness: a person's arm is weak or numb.

S: Speech problems: a person cannot speak, cannot repeat a sentence, or their speech is slurred.

T: Time to call 911. Stroke is a medical emergency and the person should get to the hospital as soon as possible.

References

Jones, J.M., and Jones, J.L., Presidential stroke: United States presidents and cerebrovascular disease, *CNS Spectrum*, 11:674-78, 2006.

Meschia, J., Safirstein, B.E., and Biller, J., Stroke and the American presidency, Journal of Stroke and Cerebrovascular Disease, 6:141-43, 1997.

National Stroke Association, http://www.stroke.org/, accessed January 7, 2016.

Why does Stephen Hawking use a wheelchair?

A

British physicist Stephen Hawking (1942–) has a neurode-generative disease called amyotrophic lateral sclerosis (ALS). He was diagnosed with ALS when he was twenty-one years old. He now relies on a wheelchair to get around and a computerized voice system to communicate. ALS is also called Lou Gehrig's disease, named for the New York Yan-kees baseball player who had the condition. Gehrig died of ALS in 1941.

People with ALS may have stiff muscles, cramping, weakness in the arms or legs, or trouble speaking or swal-lowing. These symptoms can lead to problems walking, writing, eating, or getting dressed. As the disease gets

worse, people with ALS lose the ability to walk. Most people with ALS do not experience problems with their senses, memory, or other higher cognitive functions, but in the late stages of the disease, people must use ventilators to help them breathe.

ALS attacks neurons in the motor cortex and spinal cord. These neurons send signals that control voluntary muscles and thus the ability to move. Upper motor neurons, located in the brain, send messages to other neurons in the spinal cord. Those neurons in the spinal cord (lower motor neurons) send messages to muscles. In people with ALS, these neurons die, so muscles fail to receive messages and will weaken and waste away. This causes weakness and paralysis.

The exact cause of ALS is not known. Some evidence points to a genetic problem, but other studies suggest that a virus or a neurotoxin may play a role. There is currently no cure, although drugs to treat ALS reduce the effects of motor neuron damage and ease the symptoms. For example, medicines can ease cramping, pain, or swallowing problems. Some people with ALS also benefit from physical therapy to keep them active and flexible, and others use speech therapy to improve their ability to communicate.

Although people with ALS are robbed of their ability to move, many are still able to work, travel, and have a rewarding life. Since 2014, many people have taken the ALS "ice bucket challenge" to raise awareness about ALS and contribute funding toward research to find a cure.

Do vaccines cause autism?

A

Autism spectrum disorder (autism) is a serious developmental disorder that affects approximately 1 in 68 children in the United States. Although the severity of symptoms differ from person to person, people with autism often have problems communicating with others, experience difficulties in social situations, and make repetitive motions. Genetic factors probably have a role in autism, but the overwhelming consensus of the medical and scientific community is that vaccines do not cause autism.

The U.S. Centers for Disease Control and Prevention states, "There is no link between vaccines and autism" and the World Health Organization writes, "There is no evidence of a link between MMR [measles-mumps-rubella] vaccine and autism or autistic disorders." An examination of the health records of 95,727 children, including children who were unvaccinated and some who had siblings with autism have lead researchers to conclude: "These findings indicate no harmful association between MMR vaccine receipt and ASD [autism spectrum disorder] even among children already at higher risk for ASD." These and other health organizations view vaccines as a safe and effective way to protect the public from devastating illnesses.

Many parents became concerned about their children's health after the publication of a paper in 1998 that linked the MMR vaccine to autism. This led to the "anti-vaxxer" movement, spearheaded by some celebrities, with parents who decided their children would not be vaccinated. Investigation into the 1998 study revealed serious flaws in the research and multiple subsequent studies failed to confirm the results of the original paper. In 2010, the 1998 paper was retracted from the journal (*The Lancet*) that published the study. Andrew Wakefield, the lead author of the paper, was also accused of fraud and stripped of his medical license.

References

Centers for Disease Control and Prevention, Vaccines do not cause autism, November 23, 2015, accessed March 25, 2016, http://www.cdc.gov/vaccinesafety/concerns/autism.html.

Jain, A., Marshall, J., Buikema, A., Bancroft, T., Kelly, J.P., and Newschaffer, C.J., Autism occurrence by MMR vaccine status among US children with older siblings with and without autism, *Journal of the American Medical Association*, 313:1534–40, 2015.

Wakefield, A.J., Murch, S.H., Anthony, A., Linnell, J., Casson, D.M., Malik, M., Berelowitz, M., Dhillon, A.P., Thomson, M.A., Harvey, P., Valentine, A., Davies, S.E., and Walker-Smith J.A., Ileal-lymphoid-nodular hyperplasia, non-specific colitis, and pervasive developmental disorder in children, *Lancet*, 351:637–41, 1998. Retraction, *Lancet*, 375:445, 2010.

World Health Organization, What are some of the myths—and facts—about vaccination?, March 2016, accessed March 25, 2016, http://www.who.int/features/qa/84/en/.

Why do people get Down syndrome?

A

Down syndrome is a genetic disorder that occurs when a person is born with an extra chromosome 21. The extra chromosome affects how a person develops and causes the physical characteristics of Down syndrome. In 1866, English physician John Langdon Down (1828–1896) was the first to describe the physical features common in people with the disorder that now bears his name: flattened faces, upward slanted eyes, unusually shaped ears, and wide hands. Many people with Down syndrome have heart problems, and some have digestive, hearing, and visual disorders. Intellectual problems such as trouble learning and remembering and other developmental difficulties are also common.

Approximately 1 out of every 691 babies is born with Down syndrome. The risk of having a baby with Down syndrome increases with the age of the mother. Down syndrome occurs in fewer than 15 of 10,000 babies born to mothers who are 34 years old or younger. Down syndrome occurs in about 30 of 10,000 babies born to mothers between the ages of 35 and 39 and in about 95 of 10,000 babies born to mothers 40 and older.

Although there is no cure for Down syndrome, it is possible to diagnosis the condition before a baby is born. One method for this is called amniocentesis, in which doctors check the fluid in the mother's womb for an extra chromosome 21. Tissue from the placenta and blood from the umbilical cord can also be tested for the presence of the extra chromosome.

With appropriate education programs, good health care, and a supportive home and community environment, people with Down syndrome can go to school, have jobs, and contribute to society. Speech therapy, job counseling, and physical therapy can also benefit people with Down syndrome as they navigate through life.

References

Allen, E.G., Freeman, S.B., Druschel, C., Hobbs, C.A., O'Leary, L.A., Romitti, P.A., Royle, M.H., Torfs, C.P., and Sherman, S.L., Maternal age and risk for trisomy 21 assessed by the origin of chromosome nondisjunction: a report from the Atlanta and National Down Syndrome Projects, *Human Genetics*, 125:41–52, 2009.

Mai, C.T., Kucik, J.E., Isenburg, J., Feldkamp, M.L., Marengo, L.K., Bugenske, E.M., Thorpe, P.G., Jackson, J.M., Correa, A., Rickard, R., Alverson, C.J., Kirby, R.S, and the National Birth Defects Prevention Network, Selected birth defects data from population-based birth defects surveillance programs in the United States, 2006 to 2010: featuring trisomy conditions, *Birth Defects Research Part A, Clinical and Molecular Teratology*, 97:709-25, 2013.

Parker, S.E., Mai, C.T., Canfield, M.A., Rickard, R., Wang, Y., Meyer, R. E., et al., Updated national birth prevalence estimates for selected birth defects in the United States, 2004-2006, *Birth Defects Research Part A Clinical and Molecular Teratology*, 88:1008-16, 2010.

? **?** **?** **?** **?**

Q

What is face-blindness?

A

Faces are really important to people. They tell us all kinds of information about the person who belongs to the face. Is it a man or a woman? Is he old or young? How does he feel? Is he afraid, is he threatening, is he in pain, is he aroused? Does he see me? Where is he looking? Is he telling the truth? Maybe most important—have I seen this face before? Is this a childhood friend or a neighbor, or someone I have never met? This information is not just important for finding a date, it is important for survival. For this reason humans are optimized to see faces, and in fact will see faces even where they don't exist (like the face of Mother Teresa in a cinnamon roll).

There is a part of the human brain, called the fusiform face area, dedicated to recognizing faces. We know this partly because damage to this area impairs a person's ability recognize familiar faces, even one's own face. This condition is called *prosopagnosia*. People with this disorder can still recognize other people by their voice and mannerisms, still know that they are looking at a person, and may learn to distinguish faces by intentional deconstruction of features. What is different is that faces are no more recognizable or interpretable than are inanimate objects. Because faces are visually complex, they can be very difficult to tell apart. Imagine trying to tell trees apart by the pattern of their bark: you can do it, but it isn't intuitive or natural. Interestingly, losing the ability to recognize faces does not necessarily impair one's ability to recognize facial expressions. These are very high-level skills, and visual processing in the brain is very complex.

Prosopagnosia is not always secondary to brain damage. Some people are born without the ability to recognize faces. People with congenital prosopagnosia may become very adept at compensating for their deficits, so much so that they may be surprised to find that they even have a problem. Some famous people who have put a face to congenital prosopagnosia include neurologist Oliver Sacks and artist Chuck Close.

References

Calder, A.J., and Young, A.W., Understanding the recognition of facial identity and facial expression, *Nature Reviews Neuroscience*, 6:641–51, 2005.

Kanwisher, N., and Galit, Y., The fusiform face area: a cortical region specialized for the perception of faces, *Philosophical Transactions of the Royal Society B: Biological Sciences*, 361:2109–128, 2006.

? **?** **?** **?** **?**

What causes migraines?

A

Most people are familiar with migraines because they either get them or know someone who does. Migraines affect millions of people worldwide. Anyone who has had one can tell you that to call a migraine a "bad headache" isn't doing it justice.

In addition to intense, throbbing pain, migraines are often accompanied by nausea, vomiting, extreme sensitivity to light and sound, blurred vision, fatigue, and fainting. Many people also experience visual, somatosensory, or motor disturbances—a so-called aura—during or immediately before a headache starts. Migraines typically last between 2 and 72 hours. They are substantially worse for some people than they are for others. Some migraineurs only get them once or twice a year (or less) but some people get them so frequently that they become debilitating.

What's the problem? For many years it was thought that migraines were the result of vasodilation (opening of blood vessels) in the head. In line with this assumption, the first effective treatments were drugs that constricted blood vessels. But this turned out to be a red herring, because the real source of migraines is neurological. A growing body of evidence suggests that the culprit is calcitonin gene-related peptide (CGRP), a string of amino acids produced in the nervous system. CGRP has two important functions: it is a vasodilator and it is a pain-signaling neurotransmitter. It appears that release of this substance sensitizes nerves in the face, head, and jaw as well as dilating the cerebral blood vessels. People who get migraines seem to be particularly sensitive to the effects of CGRP. Several pharmaceutical companies are now developing drugs to prevent CGRP from binding to its receptor, which could be good news for migraine sufferers. Although this is good news, there are still many outstanding questions about migraines, such as why are some people more sensitive to CGRP than others, and what causes the release of CGRP that leads to a migraine? Maybe we can think about that once the pounding pain in our heads lets up.

References

Lassen, L.H., Haderslev, P.A., Jacobsen, V.B., Iversen, H.K., Sperling, B., and Olesen, J., CGRP may play a causative role in migraine, *Cephalalgia*, 22:54–61, 2002.

Underwood, E., Will antibodies finally put an end to migraines?, http://www.sciencemag.org/news/2016/01/feature-will-antibodies-finally-put-end-migraines, accessed January 12, 2016.

What causes an "ice cream" headache?

A

Saying "sphenopalatine ganglioneuralgia," the medical term for an ice cream headache, might be enough to cause your head to throb. Ice cream headaches, also called brain freeze, happen when people eat cold foods or drink cold beverages too fast.

A common explanation for brain freeze pain involves the way blood vessels and nerves on roof of the mouth respond to cold temperatures. Cooling causes blood vessels to constrict and then expand as they warm up. Nearby pain receptors respond to these blood vessel changes and send information about pain to the brain. The nerve sending these signals is also used to send information about the head and face. Therefore, the brain interprets this painful sensory information as a headache, even though the origin of the signals is in the roof of the mouth.

Cooling of the upper palate may also cause rapid increases in the size the anterior cerebral artery in the brain. This artery brings blood to parts of the frontal and parietal lobes of the brain. When the diameter of these blood vessels gets larger, more blood flows into the brain. Because this blood has nowhere to go inside the solid skull, there is an increase in pressure inside the head that causes pain. When the blood vessels return to their normal size, the headache pain goes away.

Whatever the cause, it's a good thing ice cream headaches last only a short time.

What causes some people to twitch and swear?

A

In 1825, French neurologist Jean-Marc Itard (1774-1838) wrote about a woman who could not control her movement and suddenly burst out cursing and swearing. Decades after this report, neuropsychiatrist Georges Albert Edouard Brutus Gilles de la Tourette (1857-1904) described more patients with uncontrollable facial twitching (tics) similar to that mentioned by Itard.

Later named Tourette syndrome (TS), the movement disorder described by Tourette and Itard almost always starts in childhood. Common symptoms include eye blinking, facial tics, grunting, and involuntary head, neck, and shoulder jerks. Less commonly, some people burst out swearing or repeat what other people say. The Centers for Disease Control and Prevention estimates that 0.3 percent of all children have some form of TS with boys showing a higher incidence of the disorder than girls.

There are currently no tests that can screen whether a child will develop TS. The condition is diagnosed based on the symptoms and when they occur. Furthermore, the exact cause of TS is not known, but genetics seems to be an important factor. Fathers with TS have a 50 percent chance of passing the TS gene to their children. Even if the child has the gene for TS, he or she may not show any signs of the disorder or the symptoms may be milder or different. Also, identical twins, who both have the TS gene, may have different symptoms or one twin will have TS and the other twin will not. These observations suggest that genetics is not the only cause of TS, and other factors such as the environment can contribute to the manifestation of symptoms.

Although there are no cures for TS, the symptoms in some people improve as they get older. Some get relief from their tics using medications used to treat psychosis, attention deficit/hyperactivity disorder, and obsessive compulsive disorder or with behavioral therapy designed to teach them how to manage their tics.

Research to unravel the mystery of TS continues in laboratories around the world. This work seeks to define the causes of TS, identify the changes in brain anatomy and chemistry, and develop better treatments for people with the disorder.

References

Centers for Disease Control and Prevention, Prevalence of diagnosed Tourette syndrome in persons aged 6-17 years—United States, 2007, *Morbidity and Mortality Weekly Report*, 58:581–85, 2009

Serajee, F.J., and Mahbubul Huq, A.H., Advances in Tourette syndrome: Diagnosis and treatment, *Pediatric Clinics of North America*, 62:687–701, 2015.

What is the most common mental health issue?

A

Anxiety disorders are extremely common; in fact, they are probably the most common mental disorder. Several conditions fall under the anxiety umbrella, including generalized anxiety disorder, panic disorder, and social anxiety disorder. Like depression, anxiety disorders occur when normal brain processes run to extremes. For anxiety disorders, the problem is excessive fear.

Scientists know that the amygdala, an evolutionarily old structure found deep in the temporal lobe, is important for fear processing. It is natural to assume that abnormalities in the amygdala might be involved with anxiety disorders. This indeed seems to be the case. Specifically, it appears that there is a problem with the connection between the prefrontal cortex and the amygdala. The source of this problem is a complicated combination of genetics and environment, an explanation that is not very satisfying. Risk factors include family history of anxiety, being shy, being poor, experiencing trauma, and being female. Anxiety disorders can also be caused by drug and alcohol abuse and can be exacerbated by caffeine.

Treatments for anxiety disorders include psychotherapy and medications. Antidepressants, especially selective serotonin uptake inhibitors (SSRIs) can be helpful in treating anxiety. Benzodiazepines, which are also used to treat seizures and induce sleep, can sometimes control the symptoms of anxiety. Likewise, beta-blockers, which are frequently used to treat heart conditions, can reduce shaking and racing heartbeat in people with anxiety disorders.

As disorders go, anxiety is similar to depression, and the two conditions tend to exist together. They are both disorders of the limbic system—the brain's emotional processing machinery. This is why antidepressants can effectively treat anxiety. Taken together, these disorders represent a tremendous amount of personal suffering and a heavy economic burden for society.

References

Kasper, S., Gryglewski, G., and Lanzenberger, R., Imaging brain circuits in anxiety disorders, *Lancet Psychiatry*, 1:251–52, 2014.

National Institute of Mental Health, Anxiety disorders, March 2016, accessed March 31, 2016, http://www.nimh.nih.gov/health/topics/anxiety-disorders/index.shtml.

? ? ? ? ?
Q

What is depression?

A

Anybody can, and everybody does, feel down from time to time. This is a completely normal and adaptive reaction to the fact that life can really be a bummer. We lose our jobs, relationships end, we get sick, and people we love die. Frankly, it can be depressing. But sometimes, that low mood stops being functional and starts to be all-consuming. This is what is referred to as major depressive disorder, sometimes called clinical depression or simply depression.

The symptoms of depression include persistent sadness, hopelessness, irritability, feelings of guilt or worthlessness, loss of pleasure in hobbies, decreased libido, fatigue, sleep problems, memory problems, weight changes, aches and pains, digestive problems, and, most concerning, thoughts of suicide. Depression is clearly a very unpleasant and potentially dangerous condition. From a treatment perspective, depression poses a thorny problem. The difference between clinical depression and run-of-the-mill depression is a matter of persistence and degree. There are no lab tests for major depressive disorder, and that means it can be difficult to diagnose.

The exact cause of clinical depression is unknown, but it almost certainly arises from the interaction of multiple factors, including genetic predisposition, psychosocial situation, and hormonal state. Other disorders, both mental and physical, usually accompany depression. All of this means that each person's disease is unique, and therefore depression can be difficult to treat. The social stigma attached to depression doesn't help matters, and consequently people may suffer in silence rather than seek help.

There are a few treatment options for depression. Many people with depression undergo therapy with a psychiatrist or psychologist. Addressing the underlying issues that cause depression will sometimes resolve the problem. Antidepressant medications may be also be prescribed, although therapy with a psychiatrist or psychology is often continued. Antidepressant medications, for example, selective serotonin reuptake inhibitors (SSRIs), attempt to correct a depressed mood by changing the brain's chemistry. Some drugs work for some people, other drugs work better for others, and some people's depression is not responsive to medication at all. Unfortunately, the only way to know is to try them until one works. Medications have side effects that may or may not be tolerable. Antidepressants also take time to work, which means they won't be useful in an emergency situation. When medications fail, or when they are unsafe (in pregnancy, for example), electroconvulsive (also known as electroshock) therapy can be considered as option.

The good news is that the social stigma associated with depression is waning and more treatment options are becoming available. This gives hope to the millions of people affected by depression.

References

National Institute of Mental Health, Depression, https://www.nimh.nih.gov/health/topics/depression/index.shtml, March 2016, accessed March 29, 2016.

What is electroshock therapy and why is it used?

A

In the early 1930s, a Hungarian psychiatrist observed a seemingly inverse relationship between epilepsy and schizophrenia. On the basis of this observation he reasoned that intentionally causing seizures in patients with schizophrenia might alleviate the symptoms of their illness, and he induced seizures in some of his patients using pharmacological agents. Although his premise was incorrect, the procedure met with some success. At a time when treatments for psychiatric disorders were limited, this new intervention was rapidly adopted by physicians all over the world.

Unfortunately, the drugs used to cause seizures had serious adverse side effects. In the late 1930s, electrical stimulation was introduced as an alternative. This was the beginning of electroconvulsive therapy (ECT), which is the induction of a seizure by electrical stimulation of the brain through electrodes placed on the scalp. This therapy is still in use today, although rarely for the treatment of schizophrenia.

Many people have a strong negative reaction to the idea of ECT, but it is one of the most effective treatments for severe and treatment-resistant depression. Although it is considered safe, there are some side effects, most notably memory impairment. ECT can cause both anterograde and retrograde amnesia, which means it can disturb past memories as well as prevent the formation of new ones, although this is usually not a permanent condition. In addition, while response rates are between 60 percent and 90 percent, there is a high rate of relapse, with up to 50 percent of initial responders experiencing relapse within six months.

ECT falls into the surprisingly large category of psychiatric treatments that were accidentally discovered and that no one can satisfactorily explain. Despite having been used in clinical practice for decades, no one really knows how or why it works. It seems to work equally well for major depressive disorder and bipolar disorder and is also effective in treating catatonia and mania. It is surprising that such a dramatic and nonspecific intervention works at all, much less that it should work to treat such a large number of different disorders. The therapeutic effect may result from a reduction in neural inhibition or from new neuron growth, or both, or something else.

The current trend is away from ECT and toward alternative stimulation technologies, such as transcranial magnetic stimulation, deep brain stimulation, and epidural cortical stimulation. The hope is that some of these stimulation techniques will provide the same benefits with reduced side effects. For now, ECT is the gold standard.

Reference

Hoy, K.E., and Fitzgerald, P.B., Brain stimulation in psychiatry and its effects on cognition, *Nature Reviews Neurology*, 6:267–75, 2010.

What is a fontal lobotomy, and why was it performed?

A

A lobotomy is a neurosurgical intervention designed to treat psychiatric disorders (a psychosurgery). The objective of a lobotomy (also known as a leucotomy) is to mechanically sever the long tracts of the prefrontal cortex. Developed in 1935 by Portuguese neurologist Antonio Egas Moniz (1874-1955), the lobotomy was expected to reduce the symptoms of severe and otherwise untreatable mental disorders, such as schizophrenia or bipolar disorder.

Unfortunately, the effects of the surgery were variable and often produced profound and deleterious side effects, ranging from emotional and intellectual deficits to death. The surgery did reduce symptoms in some patients and made others more institutionally manageable. In the absence of alternative treatments, the lobotomy became a popular surgery for physicians, if not for their patients, who rarely consented to the procedure. Many people would not have met the modern criteria for competence to consent.

Moniz was awarded the Nobel Prize in Physiology or Medicine in 1949 for developing the lobotomy, but the name most strongly associated with the procedure is Walter Freeman (1895-1972). Freeman popularized the procedure in the United States and personally performed thousands of lobotomies. He also invented the transorbital or "ice pick" lobotomy that could be performed in psychiatric hospitals and asylums where surgical suites were not available. Freeman clearly had a natural flair for the dramatic and often made the surgery into a show, which is callous and unprofessional. It is also clear that even if he started with good intentions, he got carried away, performing very risky surgeries on patients, including children, when it was obviously not indicated. Partly as a result of this overexuberance, the word *lobotomy* is now synonymous with barbaric and inhumane medical treatment.

When antipsychotic medications became available in the 1950s, the lobotomy rapidly went extinct. However, psychosurgeries are still with us, serving as a treatment of last resort for people with profound and intractable illness. These descendants of the lobotomy are much more limited and specific and have better response rates and fewer, less severe side effects. Increasingly these ablative surgeries are also being phased out in favor of electrical deep brain stimulation (DBS). The mechanism of action for DBS is not well understood, which should be a red flag, but it can be modulated and reversed. For people with serious illness, it may be worthwhile.

Reference

Kopell, B.H., Machado, A.G., and Rezai, A.R., Not your father's lobotomy: psychiatric surgery revisited, *Clinical Neurosurgery*, 52:315–30, 2005.

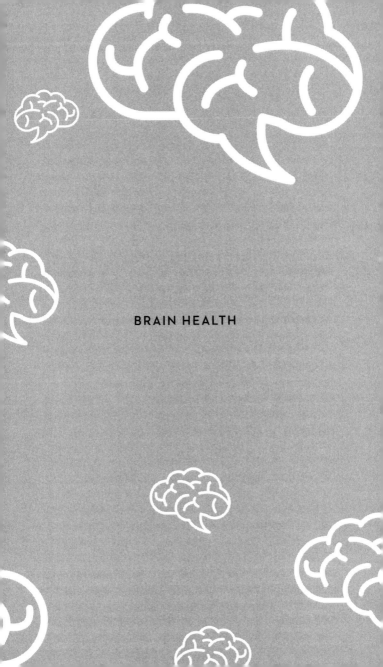

BRAIN HEALTH

What can I do to keep my brain healthy?

A

Keeping your brain in its best shape requires common sense and informed knowledge. Unlike cells in other parts of the body, nerve cells in the brain are not replaced after they are damaged. Although the brain has a tremendous capacity to adapt and rewire itself after injury, the best course of action is to prevent or slow any neurological problems before they occur.

Lifestyle choices can greatly reduce the risk of injury to your brain and may slow the start of cognitive problems later in life. For example, wear a seat belt every time you ride in a car. Car accidents cause many brain injuries, and seat belts can minimize damage if you are involved in a crash. Wearing a helmet is also a good idea when you ride a bike, skate, ski, or snowboard. Make sure the helmet meets or exceeds the American National Standards Institute and Snell Memorial Foundation standards for safety.

Adequate sleep, a balanced diet, and exercise are the three most important components of brain health. For peak mental performance, adults should get about seven hours of sleep each night. Although the amount of sleep needed by different people varies, lack of sleep can have detrimental effects on a person's memory, emotions, and physical well-being. A balanced diet will bring all of the vitamins, minerals, and fuel necessary to power your brain through the day and night. An imbalance of these materials may prevent your brain from working properly. Mental and physical exercise also contribute to good brain hygiene. Regular physical exercise appears to reduce the risk of cognitive decline that often accompanies old age. Mental exercise is also associated with lower risk of Alzheimer's disease and even gives a boost to the immune system. The most effective type and amount of physical and mental exercise is not known. "Complete two crossword puzzles, run two miles and call me in the morning." Could this be a prescription for health in the future?

What happens to the brain as it ages?

A

You probably don't really want to know, but you probably already do. As you get older, your neurons have a harder time communicating with each other. One reason for this trouble is because the insulating layer on the outside (the myelin) begins to degrade. At the same time, blood flow to the brain is reduced and cellular damage due to free radicals goes up. Perhaps most unsettling is the fact that your brain actually shrinks: you lose about 5 percent of your brain volume per decade after the age of 40. This shrinking is not evenly distributed across the brain: it seems to hit the prefrontal cortex and hippocampus the hardest. Because these brain areas are associated with memory, planning, attention, and other higher level executive functions, you would expect to see deficits in these functions—which you do—although in healthy aging, the difference is not as dramatic as you might expect.

Part of the reason the brain does so well as you age is likely because it is able to compensate: it is able to reroute traffic to avoid damaged areas. Some people are able to compensate better than others for reasons that are not totally clear but almost certainly involve a combination of genetics and environment.

Is there anything you can do to avoid this slow slide into decay? Not totally, but there are things you can do to resist. Mostly you just need to do the things that you already know you're supposed to do: exercise, eat well, don't drink too much. You can also try to engage your brain, pick up new skills, form relationships, and talk to people. Besides making you healthier, following these guidelines will make you a more well-rounded and interesting person. If your cognitive skills decline as you age, take heart in the fact that as you have more experiences, you also become wiser and more able to take things in stride.

References

Peters, R., Ageing and the brain, *Postgraduate Medical Journal*, 82:84–88, 2006.

Rodgers, A.B., Alzheimer's disease. Unraveling the mystery, https://www.nia.nih.gov/alzheimers/publication/part-1-basics-healthy-brain/changing-brain-healthy-aging, U.S. Department of Health and Human Services, 2008.

Is laughter really the best medicine?

A

Laughter probably won't hurt (unless you pull out your stitches from a recent surgery), but it may not cure your illness either. Writer Norman Cousins brought laughter to the attention of the medical community when he wrote that 10 minutes of laughing to funny movies brought him relief from pain caused by ankylosing spondylitis. But should we abandon our medicines and cancel our surgeries and instead fire up a few episodes of *The Big Bang Theory*?

Laughter does have direct effects on the body. The heart, lungs, muscles, immune system, and endocrine system are all activated when a person laughs. For example, laughter elevates blood pressure and pulse rate and provides a boost to the immune system. A good, hearty laugh can also reduce stress, put a person in a better mood, and help someone cope with a traumatic event. As Cousins pointed out, laughter can serve as an analgesic and reduce pain.

The underlying physiological mechanisms for any beneficial effects of laughter are not known. Laughing may help the respiratory system by clearing out mucus, and the cardiovascular system may be strengthened when the heart gets a workout. Stress hormones in the blood are reduced after laughing, and beneficial immune cells are boosted. But these effects may not be specific for laughter.

Any activity that stimulates the respiratory and cardiovascular system may provide the same benefits as laughter. One special characteristic of laughter is mirth, but it is unclear how this feature contributes to any health benefits. Finally, the amount of laughter necessary to result in better health outcomes and which conditions would benefit from laughter are not known. Doctors do not yet write prescriptions for comics, cartoons, or funny movies.

Reference

Cousins, N., *Anatomy of an Illness. As Perceived by the Patient* (New York: W.W. Norton, 1979).

Mora-Ripoll, R., The therapeutic value of laughter in medicine, *Alternative Therapies*, 16:56–64, 2010.

Why do my eyes hurt when I walk outside after being inside a movie theater?

A

Believe it or not, the sun really is not brighter than when you entered the theater. But it sure seems that way. To find your car, you have to squint your eyes as you stumble around the parking lot.

Remember that there are two types of photoreceptors in the retina: rods and cones. Cones need plenty of light to work and provide information about color. Rods work in low-light situations but do not send signals to the brain about color. When you first enter a dark movie theater, only cone receptors work. You are temporarily blind as you stagger to find a seat because the cone receptors aren't working in the dark and it takes rod receptors 7–10 minutes to function. When the movie starts, the colors on the screen are bright enough to activate cone receptors, but the rest of the theater is too dark for those receptors to provide any information.

After the credits roll and you step outside, the flood of sunlight makes it difficult to see. This temporary blindness occurs because your eye has adapted to the dark. The pupil of your eye, opened in the dark to let in more light, is still wide open as you exit the theater. This light stimulates the photoreceptors that have been adapted for low-light conditions inside the theater. After a few minutes outside, cone receptors start to work again and you can see normally.

Why does scratching fingernails on a chalkboard sound bad?

A

Does the sound of fingernails on a chalkboard send chills down your spine? How about the sound of Styrofoam rubbed together or scrapping metal on metal? These are some of the most unpleasant sounds rated by adults.

The key to the unpleasantness of these sounds is in the low-frequency range, because when low frequencies are filtered out of the sounds, the sound becomes less unpleasant. Their cringe-worthy nature may exist because of their evolutionary significance. Macaque monkeys have a warning cry that is similar to the sound of nails on a chalkboard. The warning call is meant to rally the troops and prepare for danger, such as an approaching predator. We humans may have retained this response from our common past primate ancestor. Other research suggests that the unpleasantness of the sound is related to the imagined physical feeling of the fingertips that would result from scrapings one's nails on a chalkboard.

The shivers sent down your spine? Neuroscientist V. S. Ramachandran hypothesizes that an evolutionary leftover from the system fish used to detect water movement may be to blame. Whatever the cause, just thinking about nails on a chalkboard is enough to cause many people to wince, flinch, and shudder.

References

Cox, T.J., Scraping sounds and disgusting noises, *Applied Acoustics*, 69:1195–204, 2008.

Halpern, L., Blake, R., and Hillenbrand, J., Psychoacoustics of a chilling sound, *Perception & Psychophysics*, 39:77–80, 1986.

Ramachandran, V.S., On the unpleasantness of certain harsh sounds, *Medical Hypothesis*, 46:487, 1996.

? **?** **?** **?**

? **?**

Why do I sneeze when I walk into bright sunlight?

A

ACHOO (autosomal-dominant compelling helio-ophthalmic outburst) syndrome, solar sneeze, photosternutatory reflex, or the photic sneeze reflex: regardless of the name, it is the phenomenon of sneezing after being exposed to bright light. Estimates vary, but 17 percent to 35 percent of the population appears to have this reflex, which causes a sneeze when they exit a dark room on a sunny day. It runs in families: if one parent sneezes to light, there is a 50 percent chance that his or her child will also have the reflex.

A regular sneeze is caused when something irritates the inner lining of the nose. The irritation activates receptors that send signals to the brain through the trigeminal nerve (the fifth cranial nerve). One theory about the cause of photic sneezes is related to the location of the optic nerve (second cranial nerve) and the trigeminal nerve. Although the optic nerve sends messages from the retina about light, the close proximity of this nerve and trigeminal nerve in photic sneezers may result in some cross-talk between nerves. Another possibility is that rapid constriction of the pupils caused by light results in mucus secretion in the nose that then leads to a sneeze. Other data suggest that the visual cortex is more sensitive in photic sneezers and that the somatosensory areas of the cortex are coactivated in these people.

Although you may think these events are harmless, they really are nothing to sneeze about if you were flying a plane or driving a car out of a tunnel on a winding road. A sneeze at the wrong time could cause temporary blindness that sends a plane on a different path or a car over a cliff. If you're a photic sneezer, think about keeping sunglasses handy for these situations.

References

Breitenbach, R.A., Swisher, P.K., Kim, M.K., and Patel, B.S., The photic sneeze reflex as a risk factor to combat pilots, *Military Medicine*, 158:806–9, 1993.

Langer, N., Beeli, G., and Jäncke, L., When the sun prickles your nose: an EEG study identifying neural bases of photic sneezing. *PLoS One*, 5.2:e9208, 2010 Feb 15. doi: 10.1371/journal.pone.0009208.

Does the full moon make people go "crazy?"

A

Surely the full moon must cause strange and unusual behavior. After all, the moon influences the ocean tides, and our bodies and brains are mostly water. Certainly weird things happen when the moon is full. The word *lunacy* comes from the Latin word *luna*, meaning moon. There must be more traffic accidents, emergency room visits, dog bites, and violence when there is a full moon.

Sorry to disappoint you. Although the moon has spawned many books, poems, and songs throughout the ages, most research has failed to find any effect of the full moon, or any other phase of the moon, on strange behavior. The data are relatively easy to find: go to the almanac, determine when the moon was in particular phases and match the number of strange events that occurred on these dates. For example, you could examine the police record over a 10-year period, count the number of murders that happened in a particular city and see if more occurred during a particular phase of the moon. You could correlate the number of reported traffic accidents, or admissions to the emergency room, or suicides or murders, or animals bites or any other unusual behavior with the phase of the moon.

These and other moon phase-behavior correlations have been studied and published in the scientific literature. Assaults, murders, arrests? Most correlations show no relationship with moon phase. Psychiatric admissions, suicides, and calls to suicide prevention lines? There are no increases around the full moon. Emergency room visits, dog bites, heart attacks, drug overdoses, nosebleeds, surgical complications? No correlation. Same for the number of traffic accidents: there are no more traffic accidents when the moon is full than in any other phase.

Even if a significant correlation between a strange behavior and the occurrence of the full moon was found, the result is at best difficult to interpret and likely meaningless. A correlation means that events vary in a predictable way; a correlation does not mean that one event causes the other to happen. So those positive correlations found in a few studies do not provide any evidence that a particular behavior was caused by the full moon.

If there is no evidence that the full moon causes strange, unusual, or dangerous behavior, why do people believe it does? First, it is interesting to note that people in stressful professions hold some of the strongest beliefs that the full moon has power over behavior: mental health workers and police officers believe most strongly that the full moon can result in unusual human behavior. Assigning blame to unfortunate, unavoidable events may help these people cope with stressful work environments. Second, people may have selective memories. When an event occurs during a full moon, people remember it; when the same type of event happens during a different moon phase, the situation is not memorable. Third, a good moon story sells. The media is eager to write up articles that fascinate the public, and a story about how crime increased on a particular day sells newspapers. Such stories also reach the desks of police officers, who may then expand law enforcement on full moon days. More police on the streets results in more arrests that reinforces that belief that the moon was involved. Which brings us to a fourth reason for the belief: self-fulfilling prophecies. If you believe something will happen, you will make it happen.

References

Rotton, J., Kelly, I.W., and Elortegui, P., Assessing belief in lunar effects: known-groups validation, *Psychological Reports*, 59:171–74, 1986.

Vance, D.E., Belief in lunar effects on human behavior, *Psychological Reports*, 76:23–24, 1995.

Wilson, J.E. II, and Tobacyk, J.J., Lunar phases and crisis center telephone calls, *Journal of Social Psychology*, 130:47–51, 1990.

Why don't woodpeckers get headaches?

A

If you beat your head against a wooden pole 20 times per second, 12,000 times a day, at a speed of 25 km/hr, you will not only have a headache, you will also suffer some serious head and brain damage. But that's what woodpeckers do without any obvious injury to their bodies. The bird's unusual anatomy appears to be the key to preventing injury and avoiding headaches.

Woodpeckers have skulls unlike those of any other bird. For starters, woodpecker skulls are thicker and denser. They also have a thin layer of cerebrospinal fluid between their brains and skulls, so their small brains sit a bit closer to the skull. This may prevent the smooth woodpecker brain from moving within the skull when it is jolted. Woodpecker skulls also have special plate-like spongy bone concentrated in front and back. The elastic nature of this bone may cushion the brain each time the bird slams its head into a tree. The woodpecker beak is also unusual: it has a longer outer tissue in the upper beak compared to the outer tissue of the lower beak, and a shorter upper beak bone compared to the lower beak bone. A beak with this configuration may redistribute impact to the head.

Perhaps the most remarkable anatomical feature of a woodpecker is its "built-in" shock absorber: a long hyoid bone. The woodpecker hyoid bone is attached to the end of the bird's tongue, then it goes through the lower jaw, divides in two, wraps around the back of the skull, travels over the top of the head, and then reattaches at the right nostril, just in front of the eyes. The location and stiffness of the hyoid bone and connecting muscles are thought to reduce the rotation of the head and neck and therefore lessen stress on the brain.

Woodpecker skulls and brains may provide insight into the design of new bicycle, motorcycle, and football helmets. Who knew bird brains could be so useful?

References

Liu, Y., Qiu, X., Zhang, X., and Yu, T.X., Response of woodpecker's head during pecking process simulated by material point method, *PLoS One*, 10.4:e0122677.. doi: 10.1371/journal.pone.0122677, 2015.

Wang, L., Cheung, J.T., Pu, F., Li, D., Zhang, M., and Fan, Y., Why do woodpeckers resist head impact injury: a biomechanical investigation, PLoS One. 6.10:e26490. doi: 10.1371/journal.pone.0026490, 2011.

APPENDICES

APPENDIX 1

Greek and Latin Origins of Neuroscientific Terminology

TABLE A.1.

Commonly Used Neuroscience Words and Their Definitions

abducens = drawing away
ablation = carrying away
acetylcholine = vinegar bile
adrenaline = near the kidney
afferent = to carry
agnosia = no knowledge
alar = wing-like
alexia = no words
alveus = canal
amacrine = no long fiber
ambidextrous = both right
ambiguus = doubtful
amblyopia = dull vision
amnesia = forgetfulness
ampulla = small bottle
amygdala = almond
analgesia = no pain
aneurysm = widening
anesthesia = no sensation
ansa = urn handle
antitoxin = against poison
aphagia = no eat
aphasia = no speech
aqueduct = water canal
arachnoid = spider web–like
arbor vitae = tree of life
arcuate = bow-shaped
astigmatism = not a point
astrocyte = star-like cell

ataxia = not orderly
atrophy = want of nourishment
auditory = to hear
aura = breath, breeze
auricle = a little ear
autonomic = self-law
axon = axis, axle
basilar = base
bouton = button
brachium = arm
bregma = front of the head
calcarine = spur-shaped
callosum = hard, tough
cannula = reed
carotid = to put to sleep
cataplexy = down stroke, seizure
catatonia = down tone
cauda = tail
cauda equina = horse tail
caudate = tail
causalgia = burning pain
cenereum = ash gray
cephalic = head
cerebellum = little brain
cerebrum = brain
ceruleus = blue
cerumen = wax
cervical = neck
chiasm = a crossing

chorea = to dance
choroid = like a membrane
ciliary = eyelash
cinereum = ashen-hued
cingulum = girdle or belt
circadian = about a day
cistern = reservoir or well
claustrum = barrier
coccyx = cuckoo
cochlea = snail shell
colliculus = little hill
conscious = aware
conus = cone or peg
coma = deep sleep
commissure = joining together
conjunctivum = join
convolution = a turn, a folding
convulsion = pulling together
cornu = horn
corona = crown
corpus = body
cortex = bark, shell, outer layer
cribiform = sieve-like
crista = crest
cruciate = cross-shaped
crus = leg
culmen = ridge, summit
cuneate = wedge
cupula = little tub
decussation = crossing
dementia = away from mind
dendrite = tree
dentate = notched
dura = hard
dyskinesia = improper motion
edema = swelling
efferent = carry out (away)
emboliform = plug-like
encephalitis = brain inflammation
encephalogram = brain writing
ependyma = upper garment
epilepsy = seizure

ethmoid = sieve-like
falx = sickle
fasciculus = little bundle
femoral = thigh
filum = thread
fimbria = fringe
fissure = cleft or slit
flocculus = tuft of wool
folium = leaf
foramen = opening
fornix = arch
fossa = trench, channel
fovea = pit, depression
fundus = bottom
funiculus = little cord
fusiform = spindle-shaped
ganglion = knot, swelling
genu = knee
geniculate = bent like a knee
glabrous = bald
gland = acorn
glia = glue
globus pallidus = pale ball
glossal = tongue
gracile = slender
gyrus = ring, circle
habenula = rein
hallucination = to wander in the mind
hemorrhage = to burst forth with blood
hippocampus = sea horse
hypnosis = sleep
hypoglossal = under the tongue
hypophysis = down growth
hypothalamus = under thalamus
incus = anvil
infundibulum = funnel
insula = island
iris = rainbow
lamina = layer, thin plate
lemniscus = woolen band or filet

lens = lentil
lenticular = shaped like a lens
limbic = border, hem, fringe
lingula = little tongue
lumbar = related to the loins
macula = spot
malleus = hammer
mater (dura) = mother
medulla = innermost, marrow
melanin = black
meninges = membrane
meningitis = membrane
inflammation
microglia = small glue
myelin = marrow
myopia = to shut eye
narcolepsy = numbness seizure
narcotic = benumbing,
deadening
neuron = nerve
node = knot
nucleus = nut
obex = barrier
oblongata = rather long
oculomotor = eye movement
oligodendrocyte = few tree cell
operculus = cover, lid
optic = for sight
oscilloscope = to examine
swinging
pallidus = pale
paralysis = to loosen
paranoia = mind beside itself
parietal = wall
patella = little plate
peduncle = stemlike
pellucidum = translucent
pia = soft
pineal = pine cone
pinna = wing
piriform = pear-shaped
pituitary = slime, mucus

placebo = to please
plexus = a braid
pons = bridge
potential = power
pterygoid = wing-shaped
pulp = flesh
pulvinar = cushion, pillow, couch
pupil = doll, little girl
putamen = shell
pyramis = pyramid
pyriform = pear-shaped
quadrigemina = four twins
rabies = to rage
ramus = branch
raphe = seam
rectus = straight
restiform = rope-like
reticular = net-like
rhodopsin = rose eye
rostral = beak
rubro = red
sacral = sacred, holy
sagittal = arrow
schizophrenia = split mind
sclera = hard
sella turcica = Turkish saddle
semilunar = half-moon
serotonin = blood-stretching
septum = wall, partition
sinus = a hollow
soma = body
somnambulism = sleep walk
somniloquism = sleep speak (talk)
spine = thorn
splenium = bandage
stapes = stirrup
stellate = star
stimulate = to goad
striated = striped
substantia nigra = black
substance
sulcus = groove, trench, furrow

synapse = connection
synesthesia = together perception
tapetum = carpet
tectum = roof
tegmentum = covering
temporal = time
tentorium = tent
thalamus = inner chamber
thoracic = chest
trabecula = little beam
tremor = to shake

trigeminal = three twins
trochlear = pulley
tubercle = swelling
uncus = hook
vagus = wandering
velum = covering
ventricle = small cavity
vermis = worm
vertex = top, summit
vesicle = blister, bladder
vitreous = glassy

TABLE A.2.

Prefixes to Commonly Used Terms in Neuroscience

PREFIX	MEANING
ab-	away from
acou-	hear
act-	do, act
ad-, aff-	to
aden-	gland
aer-	air, gas
alg-	pain
alve-	tough
ambi-	both sides
andr-	man
angi-	blood vessel, duct
ante-	before
anti-	against, counter
arachn-	spider
arch-	beginning, origin
arthr-	joint
articul-	joint
audi-	hearing
aur-	ear
ax-, axon-	axis
bar-	weight
bi-	two, double
blast-	bud
brachi-	arm
brady-	slow
capit-	head
cata-	down
caud-	tail
cell-	cell, room
centr-	center
cephal-	head
cerv-	neck
chord-	string, cord
chro-	color
chron-	time

contra-	against, counter
corpus-	body
crani-	skull
cry-	cold
cut-	skin
dendr-	tree
dent-	tooth
derm-	skin
dors-	back
dys-	bad, improper
ect-	outside
electr-	amber
encephal-	brain
end-	inside
epi-	upon, after
esthe-	feel, perceive
eu-	good, normal
exo-	outside
extra-	outside of, beyond
fibr-	fibrous, fibers
glom-	little ball
gloss-	tongue
graph-	write, scratch
hemi-	half
hyper-	above, beyond, extreme
hypno-	sleep
hypo-	below, under
inflamm-	setting on fire
infra-	beneath
inter-	among, between
intra-	inside
lingu-	tongue
medi-	middle
mega-	large
meningo-	membrane
mes-	middle
meta-	beyond
micro-	small
multi-	many
ophthalm-	eye
para-	beside, beyond
peri-	around
phag-	eat
phon-	sound
phot-	light
poly-	many
proprio-	self, one's own
resti-	rope
retro-	backward
somat-	body
spondyl -	vertebra
sub-	under, below
supra-	above
syringo-	pipe, tube
tel-	end

APPENDIX 2

Eponyms: Neuroscientific Diseases, Symptoms, Methods, and Structures Named after People

Achilles' tendon reflex
Aicardi syndrome
Alzheimer's disease
Angelman syndrome
Aqueduct of Sylvius
Arnold-Chiari malformation
Asperger syndrome
Babinski's sign
Bell's palsy
Bodian method
Broca's aphasia
Brodmann's areas
Brown-Sequard syndrome
Cajal cell
Capgras syndrome
Charcot's disease
Charcot-Marie-Tooth disease
Circle of Willis
Creutzfeldt-Jakob disease
Cushing reflex
Dandy-Walker syndrome
Down syndrome
Duchenne's dystrophy
Edinger-Westphal nucleus
Ekbom's syndrome
Foramina of Monro
Foramina of Luschka
Friedreich's ataxia
Golgi apparatus
Golgi neuron
Golgi stain
Gower's sign
Guillain-Barré syndrome
Horner's syndrome

Huntington's disease
Jacksonian epilepsy
Jendrassik's maneuver
Kluver-Bucy syndrome
Korsakoff's syndrome
Landau-Kleffner syndrome
Lewy bodies
Lewy body dementia
Lissauer's tract
Lou Gehrig's disease
Martinotti cell
Meissner's corpuscle
Merkel disc
Meynert's basal nucleus
Ménière's disease
Moro's reflex
Nissl stain
Pacinian corpuscle
Parkinson's disease
Pick's disease
Purkinje cell
Romberg sign
Schwann cell
Sylvian fissure
Tourette syndrome
Wada test
Wallenberg syndrome
Wallerian degeneration
Wernicke's aphasia
Weigert's stain
Weil stain
Wilson's disease
Zonule of Zinn

APPENDIX 3

Average Brain Weights (in Grams)

ANIMAL	BRAIN WEIGHT (G)
Sperm whale	7,800
Fin whale	6,930
Elephant	4,783
Humpback whale	4,675
Gray whale	4,317
Killer whale	5,620
Bowhead whale	2,738
Pilot whale	2,670
Bottle-nosed dolphin	1,500–1,600
Adult human	1,300–1,400
Walrus	1,020–1,126
Camel	762
Giraffe	680
Hippopotamus	582
Leopard seal	542
Horse	532
Polar bear	498
Gorilla	465–540
Cow	425–458
Chimpanzee	420
Orangutan	370
California sea lion	363
Manatee	360
Tiger	263.5
Lion	240
Grizzly bear	234
Pig	180
Jaguar	157
Sheep	140
Baboon	137
Rhesus monkey	90–97
Dog (beagle)	72
Aardvark	72
Beaver	45
Shark (great white)	34
Shark (nurse)	32
Cat	30
Porcupine	25
Squirrel monkey	22
Marmot	17
Rabbit	10–13
Platypus	9
Alligator	8.4
Squirrel	7.6
Opossum	6
Flying lemur	6
Fairy anteater	4.4
Guinea pig	4
Ring-necked pheasant	4
Hedgehog	3.35
Tree shrew	3
Fairy armadillo	2.5
Owl	2.2
Grey partridge	1.9
Rat	2
Hamster	1.4
Elephant shrew	1.3
House sparrow	1.0
European quail	0.9

Turtle.............................0.3-0.7   Goldfish...........................0.097
Bull frog...........................0.24   Green lizard......................0.08
Viper.....................................0.1

Source: Adapted from Chudler, http://faculty.washington
.edu/chudler/facts.html.

APPENDIX 4

Animal Venoms

ANIMAL	NAME OF VENOM	NEURONAL ACTION
Anemone	Stichodactyla Toxin	Blocks voltage-gated potassium channels
Anemone	ATX II	Activates voltage-gated sodium channels
Bee (honey)	Apamin	Blocks potassium channels
Bee (honey)	Tertiapin	Blocks potassium channels
Bird (pitohui)	Homobatrachotoxin	Activates sodium channels
Coral (soft)	Palytoxin	Poisons sodium/potassium pump; opens channels
Fish (pufferfish)	Tetrodotoxin (TTX)	Blocks sodium channels
Frog (poison arrow)	Batrachotoxin	Prevents sodium channels from closing
Mussel (blue)	Domoic acid	Glutamate/kainate receptor agonist
Octopus (blue-ringed)	Maculotoxin	Blocks sodium channels
Snail (marine)	Conotoxin	Several types: one blocks voltage-sensitive calcium channels; one blocks voltage-sensitive sodium channels; one blocks ACh receptors.

Snake (Eastern green mamba)	Fasciculin-I	Blocks action of acetylcholinesterase
Snake (green mamba)	Dendrotoxin	Blocks voltage-gated potassium channels
Snake (krait)	alpha-bungarotoxin	Blocks ACh (nicotinic) receptor
Snake (krait)	beta-bungarotoxin	Inhibits release of ACh at neuromuscular junction and blocks potassium channels
Snake (Eastern green mamba)	Calcicludine	Blocks voltage-gated calcium channels
Snake (black mamba)	Calciseptine	Blocks voltage-gated calcium channels
Snake (South American rattlesnake)	Crotoxin	Reduces ACh release
Snake (cobra)	Cobrotoxin	Blocks nicotinic receptors
Scorpion (Mexican)	rErgtoxin-1	Blocks potassium channels
Scorpion	Agitoxin	Blocks potassium channels
Scorpion (Central American)	rHongotoxin-1	Blocks potassium channels
Scorpion	Iberiotoxin	Blocks potassium channels
Scorpion	Margatoxin	Blocks potassium channels
Scorpion	Maurotoxin	Blocks potassium channels
Scorpion	Noxiustoxin	Blocks sodium channels

Scorpion	Kaliotoxin	Blocks potassium channels
Scorpion (South African)	Kurtoxin	Blocks calcium channels
Scorpion (Brazilian)	Tityustoxin-K	Blocks potassium channels
Scorpion	Charybdotoxin	Blocks potassium channels
Scorpion	Scyllatoxin	Blocks potassium channels
Scorpion (Indian red)	rTamapin	Blocks potassium channels
Snake (sea)	Erabutoxin	Blocks acetylcholine (nicotinic) receptors
Snake (Australian taipan)	Taicatoxin	Inhibits voltage-gated calcium channels
Snake (Australian taipan)	Taipoxin	Inhibits release of acetylcholine
Snake (Australian common brown)	Textilotoxin	Blocks release of acetylcholine
Spider (Blue Mountains funnel web)	Atracotoxin	Blocks voltage-gated calcium channels
Spider (Joro)	Joro spider toxin	Blocks glutamate receptors
Spider (African tarantula)	rStromatoxin-1	Blocks voltage-gated potassium channels
Spider (African tarantula)	SNX-482	Blocks calcium channels
Spider (banana)	Phoneutriatoxin	Slows sodium channel inactivation
Spider (Chilean fire tarantula)	Phrixotoxin	Blocks potassium channels

Spider (funnel web)	Robustotoxin	Opens sodium channels
Spider (funnel Web)	Agatoxin	Blocks calcium channels
Spider (funnel web)	Versutoxin	Opens sodium channels
Spider (Chinese bird)	HWTX-I	Blocks calcium channels
Spider (black widow)	Latrotoxin	Enhances acetylcholine release
Spider (South American rose tarantula)	Grammotoxin SIA	Blocks calcium channels
Tick (Australian paralysis)	Holocyclotoxin	Inhibits release of acetylcholine
Wasp (solitary)	Pompilidotoxin	Activates voltage-gated sodium channels
Wasp (predaceous)	Philanthotoxin	Blocks glutamate receptors

Source: Adapted from Chudler, http://faculty.washington
.edu/chudler/toxin1.html.

APPENDIX 5

Neuroscientists Awarded the Nobel Prize in Physiology or Medicine

YEAR OF AWARD	NAME(S)	BIRTH AND DEATH DATES	NATION-ALITY/ CITIZEN-SHIP	FIELD OF STUDY
1906	Camillo Golgi	7/7/1843 to 1/21/1926	Italian	Structure of the nervous system
1906	Santiago Ramon y Cajal	5/1/1852 to 10/18/1934	Spanish	Structure of the nervous system
1911	Allvar Gull-strand,	6/5/1862 to 7/28/1930	Swedish	Optics of the eye
1914	Robert Barany	5/22/1876 to 4/8/1936	Austrian	Physiology and pathology of the vestibular appa-ratus
1927	Julius Wagner-Jauregg	3/7/1857 to 9/27/1940	Austrian	Discovery of malaria inoculation to treat dementia paralytica
1932	Edgar Douglas Adrian	11/30/1889 to 8/4/1977	British	Function of neurons in sending mes-sages
1932	Sir Charles Scott Sherrington	11/27/1857 to 3/4/1952	British	Function of neurons in the brain and spinal cord
1936	Sir Henry Hallett Dale	6/9/1875 to 7/23/1968	British	Chemical trans-mission of nerve impulses
1936	Otto Loewi	6/3/1873 to 12/25/1961	German, U.S.	Chemical trans-mission of nerve impulses

1944	Joseph Erlanger	1/5/1874 to 12/15/1965	U.S.	Differentiated functions of single nerve fibers
1944	Herbert Spencer Gasser	7/5/1888 to 5/11/1963	U.S.	Differentiated functions of single nerve fibers
1949	Antonio Caetano Abreu Freire Egas Moniz	11/29/1874 to 12/13/1955	Portuguese	Leucotomy for certain psychoses
1949	Walter Rudolph Hess	3/17/1881 to 8/12/1973	Swiss	The "interbrain" (hypothalamus) used to control activity of internal organs
1957	Daniel Bovet	3/23/1907 to 4/9/1992	Italian	Work on synthetic substances that inhibit action of body substances
1961	Georg Von Bekesy	6/3/1899 to 6/13/1972	Hungarian, U.S.	Function of the cochlea
1963	Sir John Carew Eccles	1/27/1903 to 5/2/1997	Australian	Ionic mechanisms of nerve cell membrane
1963	Sir Alan Lloyd Hodgkin	2/5/1914 to 12/20/1998	British	Ionic mechanisms of nerve cell membrane
1963	Sir Andrew Fielding Huxley	12/22/1917 to 5/30/2012	British	Ionic mechanisms of nerve cell membrane
1967	Ragnar Arthur Granit	10/30/1900 to 3/12/1991	Finnish, Swedish	Mechanisms of vision, wavelength discrimination of the eye

1967	Halden Keffer Hartline	12/22/1903 to 3/17/1983	U.S.	Mechanisms of vision
1967	George Wald	11/18/1906 to 04/12/1997	U.S.	Mechanisms of vision, chemical processes
1970	Julius Axelrod	5/30/1912 to 12/29/2004	U.S.	Humoral transmitters in sympathetic nerves
1970	Sir Bernard Katz	3/26/1911 to 4/20/2003	German, British	Release of neurotransmitters from nerve terminals
1970	Ulf Svante von Euler	2/7/1905 to 3/10/1983	Swedish	Humoral transmitters in sympathetic nerves
1973	Konrad Zacharias Lorenz	11/7/1903 to 2/27/1989	Austrian	Ethology
1973	Nikolaas Tinbergen	4/15/1907 to 12/21/1988	Dutch	Ethology
1973	Karl von Frisch	11/20/1886 to 6/12/1982	Austrian	Ethology
1976	Baruch S. Blumberg	7/28/1925 to 4/5/2011	U.S.	Mechanisms for origin and dissemination of infection disease
1976	Daniel Carleton Gajdusek	9/9/1923 to 12/12/2008	U.S.	Mechanisms for origin and dissemination of infection disease
1977	Roger Guillemin	1/11/1924 to	French, U.S.	Production of peptides in the brain
1977	Andrew Victor Schally	11/30/1926 to	Polish, Canadian, U.S.	Production of peptides in the brain

1979	Allan MacLeod Cormack	2/23/1924 to 5/7/1998	South African, U.S.	Invention of computer-assisted tomography
1979	Sir Godfrey Newbold Hounsfield	8/28/1919 to 8/12/2004	British	Invention of computer-assisted tomography
1981	David Hunter Hubel	2/27/1926 to 9/22/2013	Canadian, U.S.	Information processing in the visual system
1981	Roger Wolcott Sperry	8/20/1913 to 4/17/1994	U.S.	Functions of the right and left hemispheres of the brain
1981	Torsten N. Wiesel	6/3/1924 to	Swedish, U.S.	Information processing in the visual system
1982	Bengt Ingemar Samuelsson	5/21/1934 to	Swedish	Discovery of prostaglandins
1982	John Robert Vane	3/29/1927 to 11/19/2004	British	Discovery of prostaglandins
1982	Sune K. Bergstrom	1/10/1916 to 8/15/2004	Swedish	Discovery of prostaglandins
1986	Stanley Cohen	12/17/1922 to	U.S.	Control of nerve cell growth
1986	Rita Levi-Montalcini	4/22/1909 to 12/30/2012	Italian, U.S.	Control of nerve cell growth
1991	Erwin Neher	3/20/1944 to	German	Function of single ion channels in cells
1991	Bert Sakmann	6/12/1942 to	German	Function of single ion channels in cells
1994	Alfred G. Gilman	7/1/1941 to	U.S.	Discovery of G-protein coupled receptors and their role in signal transduction

1994	Martin Rodbell	12/1/1925 to 12/7/1998	U.S.	Discovery of G-protein coupled receptors and their role in signal transduction
1997	Stanley B. Prusiner	5/28/1942 to	U.S.	Discovery of prions; a new biological principle of infection
2000	Arvid Carlsson	1/25/1923 to	Swedish	Signal transduction in the nervous system/dopamine
2000	Paul Greengard	12/11/1925 to	U.S.	Signal transduction in the nervous system
2000	Eric R. Kandel	11/7/1929 to	U.S.	Signal transduction in the nervous system/learning
2003	Paul C. Lauterbur	5/6/1929 to 3/27/2007	U.S.	Discoveries concerning magnetic resonance imaging
2003	Sir Peter Mansfield	10/9/1933 to	British	Discoveries concerning magnetic resonance imaging
2003	Roderick MacKinnon	2/16/1956 to	U.S.	Structural and mechanistic studies of ion channels
2004	Linda B. Buck	1/29/1947 to	U.S.	Discovery of odorant receptors and the organization of the olfactory system
2004	Richard Axel	7/2/1946 to	U.S.	Discovery of odorant receptors and the organization of the olfactory system

2013	James E. Rothman	11/3/1950 to	U.S.	Discovery of the machinery regulating vesicle traffic, a major transport system in our cells
2013	Randy W. Schekman	12/30/1946 to	U.S.	Discovery of the machinery regulating vesicle traffic, a major transport system in our cells
2013	Thomas C. Sudhof	12/22/1955 to	German, U.S.	Discovery of the machinery regulating vesicle traffic, a major transport system in our cells
2014	John O'Keefe	11/18/1939 to	U.S., British	Discovery of cells that constitute a positioning system in the brain
2014	Edvard I. Moser	4/27/1962 to	Norwegian	Discovery of cells that constitute a positioning system in the brain
2014	May-Britt Moser	1/4/1963 to	Norwegian	Discovery of cells that constitute a positioning system in the brain

APPENDIX 6

Milestones in Neuroscience Research

ca. 4000 B.C. Euphoriant effect of poppy plant reported in Sumerian records

ca. 2700 B.C. Shen Nung originates acupuncture

ca. 1700 B.C. Edwin Smith surgical papyrus: first written record about the nervous system

ca. 500 B.C. Alcmaion of Crotona dissects sensory nerves and describes the optic nerve

ca. 500 B.C. Empedocles suggests that "visual rays" cause sight

460–379 B.C. Hippocrates states that the brain is involved with sensation and is the seat of intelligence

387 B.C. Plato teaches at Athens; believes brain is seat of mental process

335 B.C. Aristotle writes about sleep; believes heart is seat of mental process

335–280 B.C. Herophilus (the "Father of Anatomy"); believes ventricles are seat of human intelligence

280 B.C. Erasistratus of Chios notes divisions of the brain

177 C.E. Galen lecture "On the Brain"

ca. 100 Marinus describes the tenth cranial nerve

ca. 100 Rufus of Ephesus describes and names the optic chiasm

ca. 390 Nemesius develops the doctrine of the ventricular localization of all mental functions

ca. 900 Rhazes describes seven cranial nerves and 31 spinal nerves in "Kitab al-Hawi Fi Al Tibb"

ca. 1000 Alhazen compares the eye to a camera-like device

ca. 1000 Al-Zahrawi (also known as Abulcasis or Albucasis) describes several surgical treatments for neurological disorders

1021 Ibn Al-Haytham (Alhazen) publishes *Book of Optics*

1025 Avicenna writes about vision and the eye in *The Canon of Medicine*

1284 Salvino D'Armate constructs eyeglasses

1402 St. Mary of Bethlehem Hospital is used exclusively for the mentally ill

1410 Institution for the mentally ill established in Valencia, Spain

1504 Leonardo da Vinci produces wax cast of human ventricles

1536 Nicolo Massa describes the cerebrospinal fluid

1538 Andreas Vesalius publishes *Tabulae Anatomicae*

1543 Andreas Vesalius publishes *On the Workings of the Human Body*

1543 Andreas Vesalius discusses the pineal gland and draws the corpus striatum

1549 Jason Pratensis publishes *De Cerebri Morbis*, an early book devoted to neurological disease

1550 Andreas Vesalius describes hydrocephalus

1550 Bartolomeo Eustachio describes the brain origin of the optic nerves

1558 Giambattista della Porta describes wooden hearing aids in his book *Natural Magick*

1561 Gabriele Falloppio publishes *Observationes Anatomicae* and describes some of the cranial nerves.

1562 Bartolomeo Eustachio publishes *The Examination of the Organ of Hearing*

1564 Giulio Cesare Aranzi coins the term *hippocampus*

1573 Constanzo Varolio names the pons and is the first to cut the brain starting at its base

1573 Girolamo Mercuriali writes *De nervis opticis* to describe optic nerve anatomy

1583 Felix Platter states that the lens only focuses light and the retina is where images are formed

1583 Georg Bartisch publishes *Ophthalmodouleia: das ist Augendienst* with drawings of the eye

1586 A. Piccolomini distinguishes between cortex and white matter

1587 Guilio Cesare Aranzi describes ventricles and the hippocampus

1590 Zacharias Janssen invents the compound microscope

1596 Sir Walter Raleigh mentions arrow poison in his book *Discovery of the Large, Rich and Beautiful Empire of Guiana*

1601 Hieronymus Fabricius ab Aquapendente publishes *Tractatus de Oculo Visusque Organo*, describing the correct location of the lens relative to the iris

1604 Johannes Kepler describes inverted retinal image

1609 J. Casserio publishes first description of mammillary bodies

1611 Lazarus Riverius textbook describing impairments on consciousness published

1621 Robert Burton publishes *The Anatomy of Melancholy* about depression

1623 Benito Daca de Valdes publishes the first book on vision testing and eyeglass-fitting

1627 William Harvey demonstrates a role of the brain in frog movement

1641 Franciscus de la Boe Sylvius describes fissure on the lateral surface of the brain (Sylvian fissure)

1644 Giovanni Battista Odierna describes the microscopic appearance of the fly eye in *L'Occhio della Mosca*

1649 René Descartes describes pineal as control center of body and mind

1650 Franciscus de la Boe Sylvius describes a narrow passage between the third and fourth ventricles (the aqueduct of Sylvius)

1658 Johann Jakof Wepfer theorizes that a broken brain blood vessel may cause apoplexy (stroke)

1661 Thomas Willis describes a case of meningitis

1662 René Descartes's book *De homine* is posthumously published

1664 Thomas Willis publishes *Cerebri anatome* (in Latin)

1664 Thomas Willis describes the 11th cranial nerve (accessory nerve)

1664 Thomas Willis suggests that cerebrospinal fluid is produced by the choroid plexus

1664 Gerardus Blasius discovers and names the *arachnoid*

1664 Jan Swammerdam causes frog muscle contraction by mechanical stimulation of nerve

1665 Robert Hooke details his first microscope

1667 Robert Hooke publishes *Micrographia*

1668 L'Abbe Edme Mariotte discovers the blind spot

1670 William Molins names the trochlear nerve

1673 Joseph DuVerney uses experimental ablation technique in pigeons

1681 English edition of Thomas Willis's *Cerebri anatome* is published

1681 Thomas Willis coins the term *neurology*

1684 Raymond Vieussens publishes *Neurographia Universalis*

1684 Raymond Vieussens uses boiling oil to harden the brain

1686 Thomas Sydenham describes a form of chorea in children and young adults

1695 Humphrey Ridley describes the restiform body and publishes *The Anatomy of the Brain*

1696 John Locke writes "Essay Concerning Human Understanding"

1697 Joseph G. Duverney introduces the term *brachial plexus*

1704 Antonio Valsalva publishes *On the Human Ear*

1705 Antonio Pacchioni describes arachnoid granulations

1709 Domenico Mistichelli describes the pyramidal decussation

1709 George Berkeley publishes *New Theory of Vision*

1717 Antony van Leeuwenhoek describes nerve fiber in cross section

1721 The word *anesthesia* first appears in English (in *Dictionary Britannicum*)

1736 Jean Astruc coins the term *reflex*

1740 Emanuel Swedenborg publishes *Oeconomia regni animalis*

1749 David Hartley publishes *Observations of Man*, the first English work using the word *psychology*

1750 Jacques Daviel performs the first cataract extraction on a living human eye

1752 The Society of Friends establishes a hospital-based environment for the mentally ill in Philadelphia

1755 J. B. Le Roy uses electroconvulsive therapy for mental illness

1760 Arne-Charles Lorry demonstrates that damage to the cerebellum affects motor coordination

1764 Domenico F. A. Cotugno describes spinal subarachnoid cerebrospinal fluid; shows that ventricular and spinal fluids are connected

1764 The interventricular foramen (foramen of Monroe) is named after Alexander Monroe; it was described earlier by Vieussens

1766 Albrecht von Haller provides scientific description of the cerebrospinal fluid

1772 John Walsh conducts experiments on torpedo (electric) fish

1773 John Fothergill describes trigeminal neuralgia (tic douloureux, Fothergill's syndrome)

1773 Sir Joseph Priestley discovers nitrous oxide

1774 Franz Anton Mesmer introduces "animal magnetism" (later called hypnosis)

1776 M. V. G. Malacarne publishes first book solely devoted to the cerebellum

1777 Philip Meckel proposes that the inner ear is filled with fluid, not air

1778 Samuel Thomas von Soemmerring presents the modern classification of the 12 cranial nerves

1779 Antonius Scarpa describes Scarpa's ganglion of the vestibular system

1780 Etienne Bonnot de Condillac publishes the first figure of "reflex action"

1781 Felice Fontana describes the microscopic features of axoplasm from an axon

1782 Francesco Gennari publishes work on "lineola albidior" (later known as the stripe of Gennari)

1782 Francesco Buzzi identifies the fovea

1783 Alexander Monro describes the foramen of Monro

1784 Benjamin Rush writes that alcohol can be an addictive drug

1786 Felix Vicq d'Azyr discovers the locus coeruleus

1786 Samuel Thomas Sommering describes the optic chiasm

1786 Georg Joseph Beer founds the first eye hospital in Vienna

1790 Johannes Ehrenritter describes the glossopharygeal nerve ganglion

1791 Luigi Galvani publishes work on electrical stimulation of frog nerves

1791 Samuel Thomas von Soemmering names the macula lutea of the retina

1792 Giovanni Valentino Mattia Fabbroni suggests that nerve action involves both chemical and physical factors

1796 Johann Christian Reil describes the insula (island of Reil)

1798 John Dalton, who was red-green colorblind, provides a scientific description of color blindness

1800 Alessandro Volta invents the wet cell battery

1800 Humphrey Davy synthesizes nitrous oxide

1800 Samuel von Sommering identifies black material in the midbrain and calls it the substantia nigra

1801 Thomas Young describes astigmatism

1801 Adam Friedrich Wilhelm Serturner crystalizes opium and obtains morphine

1801 Philippe Pinel publishes *A Treatise on Insanity*

1802 Thomas Young suggests the three types of retinal receptors are sufficient for color vision

1808 Franz Joseph Gall publishes work on phrenology

1809 Johann Christian Reil uses alcohol to harden the brain

1809 Luigi Rolando uses galvanic current to stimulate the cerebral cortex

1811 Julien Jean Legallois discovers a respiratory center in the medulla

1811 Charles Bell discusses functional differences between dorsal and ventral roots of the spinal cord

1812 Benjamin Rush publishes *Medical Inquiries and Observations upon the Diseases of the Mind*

1813 Felix Vicq d'Azyr discovers the claustrum

1817 James Parkinson publishes "An Essay on the Shaking Palsy"

1820 Galvanometer invented by Johann Schweigger

1821 Charles Bell describes facial paralysis ipsilateral to facial nerve lesion (Bell's palsy)

1821 Francois Magendie discusses functional differences between dorsal and ventral roots of the spinal cord

1822 Friedrich Burdach names the cingular gyrus

1822 Friedrich Burdach distinguishes lateral and medial geniculate

1823 Marie-Jean-Pierre Flourens states that cerebellum regulates motor activity

1824 John C. Caldwell publishes *Elements of Phrenology*

1824 Marie-Jean-Pierre Flourens details ablation to study behavior

1824 F. Magendie provides first evidence of cerebellum role in equilibration

1825 John P. Harrison first argues against phrenology

1825 Jean-Baptiste Bouillaud presents cases of loss of speech after frontal lesions

1825 Robert B. Todd discusses the role of the cerebral cortex in mentation, corpus striatum in movement, and midbrain in emotion

1825 Luigi Rolando describes the sulcus that separates the precentral and postcentral gyri

1826 Johannes Muller publishes theory of specific nerve energies

1827 E. Merck & Company market morphine

1832 Justus von Liebig discovers chloral hydrate

1832 Jean-Pierre Robiquet isolates codeine

1832 Massachusetts establishes a State Lunatic Hospital for the mentally ill

1832 Sir Charles Wheatstone invents the stereoscope

1833 Philipp L. Geiger isolates atropine

1834 Ernst Heinrich Weber publishes theory of "just noticeable difference" or "Weber's law"

1836 Marc Dax delivers paper on left hemisphere damage effects on speech

1836 Gabriel Gustav Valentin identifies neuron nucleus and nucleolus

1836 Robert Remak describes myelinated and unmyelinated axons

1837 Jan Purkyne (Purkinje) describes cerebellar cells; identifies neuron nucleus and processes

1837 The American Physiological Society is founded

1838 Robert Remak suggests that nerve fiber and nerve cell are joined

1838 Theodor Schwann describes the myelin-forming cell in the peripheral nervous system (Schwann cell)

1838 Jean-Etienne-Dominique Esquirol publishes *Des Maladies Mentales*, possibly the first modern work about mental disorders

1838 Napoleonic code leads to the requirement of facilities for the mentally ill

1838 Eduard Zeis publishes study about dreams in people who are blind

1839 Theodor Schwann proposes the cell theory

1839 C. Chevalier coins the term *microtome*

1839 Francois Leuret names the Rolandic sulcus for Luigi Rolando

1840 Filippo Pacini describes the Pacinian corpuscle

1840 Moritz Heinrich Romberg describes a test for conscious proprioception (Romberg test)

1840 Adolph Hannover uses chromic acid to harden nervous tissue

1840 Jules Gabriel Francois Baillarger discusses the connections between white and gray matter of cerebral cortex

1840 Adolphe Hannover discovers the ganglion cells of the retina

1841 Dorothea Dix investigates brutality within mental hospitals in the United States

1842 Benedikt Stilling is first to study spinal cord in serial sections

1842 Crawford W. Long uses ether on humans

1842 Francois Magendie describes the median opening in the roof of the fourth ventricle (foramen of Magendie)

1843 James Braid coins the term *hypnosis*

1844 Robert Remak provides first illustration of six-layered cortex

1844 Horace Wells uses nitrous oxide during a tooth extraction

1845 Ernst Heinrich Weber and Edward Weber discover that stimulation of the vagus nerve inhibits the heart

1846 William Morton demonstrates ether anesthesia at Massachusetts General Hospital

1847 Chloroform anesthesia used by James Young Simpson

1847 Chloroform anesthesia used by Marie Jean Pierre Flourens

1848 Phineas Gage has his brain pierced by an iron rod

1848 Richard Owen coins the word *notochord*

1849 Hermann von Helmholtz measures the speed of frog nerve impulses

1850 Augustus Waller describes appearance of degenerating nerve fibers

1850 Marshall Hall coins the term *spinal shock*

1850 Emil Du Bois-Reymond invents nerve galvanometer

1851 Jacob Augustus Lockhart Clarke describes the nucleus dorsalis, an area in the intermediate zone of the spinal cord gray matter

1851 Heinrich Muller is first to describe the colored pigments in the retina

1851 Marchese Alfonso Corti describes the cochlear receptor organ in the inner ear (organ of Corti)

1851 Hermann von Helmholtz invents ophthalmoscope

1851 Andrea Verga describes the cavum vergae

1852 A. Kolliker describes how motor nerves originate from the neurons in the anterior horn of the spinal cord

1852 George Meissner and Rudolf Wagner describe encapsulated nerve endings later known as Meissner's corpuscles

1853 William Benjamin Carpenter proposes "sensory ganglion" (thalamus) as seat of consciousness

1854 Louis P. Gratiolet describes convolutions of the cerebral cortex

1855 Bartolomeo Panizza shows the occipital lobe is essential for vision

1855 Richard Heschl describes the transverse gyri in the temporal lobe (Heschl's gyri)

1856 Albrecht von Graefe describes homonymous hemianopia

1857 Charles Locock observes the anticonvulsive effects of potassium bromide

1858 Joseph von Gerlach stains brain tissue with a carmine solution

1859 Rudolph Virchow coins the term *neuroglia*

1860 Albert Niemann purifies cocaine

1860 Gustav Theodor Fechner develops Fechner's law

1860 Karl L. Kahlbaum describes and names catatonia

1861 Paul Broca discusses cortical localization

1861 T. H. Huxley coins the term *calcarine sulcus*

1862 William Withey Gull describes clinical signs of syringomyelia

1862 Hermann Snellen invents the eye chart with letters to test vision

1863 Foramen of Luschka named after Hubert von Luschka

1863 Ivan Mikhalovich Sechenov publishes *Reflexes of the Brain*

1863 Nikolaus Friedreich describes a progressive hereditary degenerative CNS disorder (Friedreich's ataxia)

1864 John Hughlings Jackson writes on loss of speech after brain injury

1865 Otto Friedrich Karl Deiters differentiates dendrites and axons

1865 Otto Friedrich Karl Deiters describes the lateral vestibular nucleus (Deiters nucleus)

1866 John Langdon Haydon Down publishes work on congenital "idiots"

1866 Julius Bernstein hypothesized that a nerve impulse is a "wave of negativity"

1866 Leopold August Besser coins the term *Purkinje cells*

1867 Hermann von Helmholtz publishes *Handbook of Physiological Optics*

1867 Theodore Meynert performs histologic analysis of cerebral cortex

1868 Julius Bernstein measures the time course of the action potential

1868 Friedrich Goll describes the fasciculus gracilis

1869 Francis Galton claims that intelligence is inherited; publishes *Hereditary Genius*

1869 Johann Friedrich Horner describes eye disorder (small pupil, droopy eyelid) later to be called Horner's syndrome

1870 Eduard Hitzig and Gustav Fritsch discover cortical motor area of dog using electrical stimulation

1870 Ernst von Bergmann writes first textbook on nervous system surgery

1871 Gustav Fechner publishes work about synesthesia

1871 Silas Weir Mitchell provides detailed account of phantom limb syndrome

1872 George Huntington describes symptoms of a hereditary chorea

1872 Sir William Turner describes the interparietal sulcus

1872 Charles Darwin publishes *The Expression of Emotions in Man and Animals*

1872 Silas Weir Mitchell provides a clinical description of phantom limb pain

1873 Camillo Golgi publishes first work on the silver nitrate method

1874 Jean Martin Charcot describes amyotrophic lateral sclerosis

1874 Vladimir Alekseyevich Betz publishes work on giant pyramidal cells

1874 Roberts Bartholow electrically stimulates human cortical tissue

1874 Carl Wernicke publishes *Der Aphasische Symptomencomplex* about aphasias

1875 Sir David Ferrier describes different parts of monkey motor cortex

1875 Richard Caton is first to record electrical activity from the brain

1875 Wilhelm Heinrich Erb and Carl Friedrich Otto Westphal describe the knee-jerk reflex

1876 David Ferrier publishes *The Functions of the Brain*

1876 Franz Christian Boll discovers rhodopsin

1876 Francis Galton uses the phrase "nature and nurture" to explain heredity and environment

1877 Jean-Martin Charcot publishes *Lectures on the Diseases of the Nervous System*

1878 W. Bevan Lewis publishes work on giant pyramidal cells of human precentral gyrus

1878 Claude Bernard describes nerve/muscle blocking action of curare

1878 The first Ph.D. with *psychology* in its title is given to Granville Stanley Hall at Harvard University

1878 Paul Broca publishes work on the great limbic lobe

1878 W. R. Gowers publishes *Unilateral Gunshot Injury to the Spinal Cord*

1878 Louis-Antoine Ranvier describes regular interruptions in the myelin sheath (nodes of Ranvier)

1876 David Ferrier publishes *The Localization of Cerebral Disease*

1879 Camillo Golgi describes the musculo-tendineous organs (later known as the Golgi tendon organs)

1879 Mathias Duval introduces an improved method of embedding tissue using collodion

1879 Hermann Munk presents detailed anatomy of the optic chiasm

1879 William Crookes invents the cathode ray tube

1879 Wilhelm Wundt sets up lab devoted to study human behavior

1879 Scottish surgeon William Macewen performs successful surgery to treat a brain abscess

1880 Jean Baptiste Edouard Gelineau introduces the word *narcolepsy*

1880 Friedrich Sigmund Merkel describes free nerve endings later known as Merkel's corpuscles

1880 Thomas Graydon invents the Dentaphone, a bone conduction hearing device

1881 Hermann Munk reports on visual abnormalities after occipital lobe ablation in dogs

1883 Sir Victor Horsley describes effects of nitrous oxide anesthesia

1883 Emil Kraepelin coins the terms *neuroses* and *psychoses*

1883 George John Romanes coins the term *comparative psychology*

1884 Franz Nissl describes the granular endoplasmic reticulum (Nissl substance)

1884 Karl Koller discovers anesthetic properties of cocaine by testing it on his own eye

1884 Georges Gilles de la Tourette describes several movement disorders

1884 Theodor Meynert publishes *A Clinical Treatise on the Diseases of the Forebrain*

1884 English surgeon Richman John Godlee performs surgery to remove a brain tumor

1885 Paul Ehrlich notes that intravenous dye does not stain brain tissue

1885 Carl Weigert introduces hematoxylin to stain myelin

1885 Ludwig Edinger describes nucleus that will be known as the Edinger-Westphal nucleus

1885 Hermann Ebbinghaus publishes *On Memory*

1885 Louis Pasteur successfully vaccinates a boy who was bitten by a rabid dog

1886 Joseph Jastrow earns the first Ph.D. from the first formal program in psychology at Johns Hopkins University

1886 V. Marchi publishes procedure to stain degenerating myelin

1887 Sergei Korsakoff describes symptoms characteristic in alcoholics

1887 Alfred Binet and C. Fere publish *Animal Magnetism*, a study about hypnosis

1887 Adolf Eugen Fick makes the first contact lens out of glass for vision correction

1887 G. Stanley Hall publishes the first issue of the *American Journal of Psychology*

1887 English surgeon Victor Horsley successfully removes a spinal cord tumor

1888 William Gill describes anorexia nervosa

1888 William W. Keen Jr. is first U.S. doctor to remove intracranial meningioma

1888 Hans Chiari introduces the term *syringomyelia*

1888 Giovanni Martinotti describes cortical cells later known as Martinotti cells

1889 Santiago Ramon y Cajal argues that nerve cells are independent elements

1889 William His coins the term *dendrite*

1889 Sir Victor Horsley publishes somatotopic map of monkey motor cortex

1889 Carlo Martinotti describes cortical neuron with ascending axon (this neuron now bears his name, Martinotti cell)

1889 F. C. Muller-Lyer discovers the Muller-Lyer illusion

1890 Wilhelm Ostwald develops the membrane theory of nerve conduction

1890 William James publishes *Principles of Psychology*

1890 The term *mental tests* was coined by James Cattell

1891 Wilhelm von Waldeyer coins the term *neuron*

1891 Luigi Luciani publishes manuscript on the cerebellum

1891 Heinrich Quinke develops the lumbar puncture (spinal tap)

1892 Santiago Ramon y Cajal publishes *Structure of the Retina*

1892 Salomen Eberhard Henschen localizes vision to calcarine fissure

1892 Arnold Pick first describes Pick's disease

1893 Paul Emil Flechsig describes myelinization of the brain

1893 Charles Scott Sherrington coins the term *proprioceptive*

1894 Franz Nissl stains neurons with dahlia violet

1895 William His first uses the term *hypothalamus*

1895 Wilhelm Konrad Roentgen invents the X-ray

1895 Heinrick Quincke performs lumbar puncture to study cerebrospinal fluid

1895 Formalization of the cranial nerve number system published in *Basle Nomina Anatomica*

1896 Max von Frey details stimulus hairs to test the somatosensory system

1896 Rudolph Albert von Kolliker coins the term *axon*

1896 Camillo Golgi discovers the Golgi apparatus

1896 Joseph Babinski describes the Babinski sign

1896 Emil Kraeplein describes dementia praecox

1896 John William Strutt publishes *The Theory of Sound*

1897 Karl Ferdinand Braun invents the oscilloscope

1897 John Jacob Abel isolates adrenaline

1897 Charles Scott Sherrington coins the term *synapse*

1897 Ferdinand Blum uses formaldehyde as brain fixative

1897 Acetylsalicylic acid (aspirin) is synthesized by Felix Hoffmann

1898 Charles Scott Sherrington describes decerebrate rigidity in cat

1898 Edward Lee Thorndike describes the "puzzle box"

1898 Bayer Drug Company markets heroin as nonaddicting cough medicine

1898 John Newport Langley coins the term *autonomic nervous system*

1898 Angelo Ruffini describes encapsulated nerve endings later known as Ruffini corpuscles

1899 Francis Gotch describes a "refractory phase" between nerve impulses

1899 Miller Hutchison invents the one of the first electric hearing aids, called the Akoulalion

1899 Karl Gustav August Bier uses cocaine for intraspinal anesthesia

1900 Sigmund Freud publishes *The Interpretation of Dreams*

1900 Charles Scott Sherrington states that cerebellum is head ganglion of the proprioceptive system

1900 M. Lewandowsky coins the term *blood-brain barrier*

1902 Julius Bernstein proposes membrane theory for cells

1902 Oskar Vogt and Cecile Vogt coin the term *neurophysiology*

1903 Ivan Pavlov coins the term *conditioned reflex*

1903 Alfred Walter Campbell studies cytoarchitecture of anthropoid cerebral cortex

1904 Procaine is synthesized

1904 Thomas Elliott suggests that autonomic nerves may release chemical transmitters

1905 Alfred Binet and Theodore Simon develop the first intelligence test

1905 John Newport Langley coins the phrase *parasympathetic nervous system*

1905 Austrian ophthalmologist Eduard Zinn performs the first successful human corneal transplant

1906 Alois Alzheimer describes presenile degeneration

1906 Sir Charles Scott Sherrington publishes *The Integrative Action of the Nervous System*, which describes the synapse and motor cortex

1907 Ross Granville Harrison describes tissue culture methods

1907 John Newport Langley introduces the concept of receptor molecules

1908 Vladimir Bekhterew describes the superior nucleus of the vestibular nerve (Bekhterew's nucleus)

1908 Victor Alexander Haden Horsley and Robert Henry Clarke design a stereotaxic instrument

1908 Willem Einthoven makes string galvanometer recordings from the vagus nerve

1908 Oberga introduces the cisterna puncture, a method to access the cerebrospinal fluid through the cistena magna

1909 Tetrodotoxin isolated from the pufferfish and named by Yoshizumi Tahara

1909 Harvey Cushing is first to electrically stimulate human sensory cortex

1909 Korbinian Brodmann describes 52 discrete cortical areas

1909 Karl Jaspers publishes *General Mental Illness*

1910 Emil Kraepelin names Alzheimer's disease

1911 Eugen Bleuler coins the term *schizophrenia*

1911 George Barger and Henry Dale discover norepinephrine (noradrenaline)

1912 Original formula for the intelligence quotient (IQ) developed by William Stern

1912 Phenobarbital brought to market

1913 Santiago Ramon y Cajal develops gold chloride–mercury stain to show astrocytes

1913 Edwin Ellen Goldmann finds blood-brain barrier impermeable to large molecules

1913 Edgar Douglas Adrian publishes work on all-or-none principle in nerve

1913 Walter Samuel Hunter devises delayed-response test

1914 Henry H. Dale isolates acetylcholine

1915 J. G. Dusser De Barenne describes activity of brain after strychnine application

1916 Richard Henneberg coins the term *cataplexy*

1916 George Guillain, Jean Alexander Barré, and Andre Strohl describe an acute inflammatory demyelinating polyneuropathy (Guillain-Barré syndrome)

1916 Shinobu Ishihara publishes a set of plates to test color vision

1918 Walter E. Dandy introduces ventriculography

1919 Cecile Vogt describes more than 200 cortical areas

1919 Walter E. Dandy introduces the air encephalography

1919 Gordon Morgan Holmes localizes vision to striate area

1919 Pio del Rio Hortega divides neuroglia into microglia and oligodendroglia

1919 Konstantin Tretiakoff describes changes in the substatia nigra in people with Parkinson's disease

1920 Henry Head publishes *Studies in Neurology*

1920 Stephen Walter Ranson demonstrates connections between the hypothalamus and pituitary

1920 John B. Watson and Rosalie Rayner publish experiments about classical conditioning of fear (Little Albert experiments)

1921 Otto Loewi publishes work on *Vagusstoff*

1921 Hermann Rorschach develops the inkblot test

1921 John Augustus Larsen and Leonard Keeler develop the polygraph

1921 del Rio Hortega describes microglia

1923 Capgras syndrome described by Joseph Capgras

1924 Charles Scott Sherrington discovers the stretch reflex

1925 C. von Economo and G. N. Koskinas revise Brodmann's cortical nomenclature of the cerebral cortex

1927 Chester William Darrow studies galvanic skin reflex

1928 Philip Bard suggests the neural mechanism of rage is in the diencephalon

1928 Walter Rudolph Hess reports "affective responses" to hypothalamic stimulation

1928 John Fulton publishes his observations (made in 1926 and 1928) of the sounds of blood flowing over the human visual cortex

1929 Hans Berger publishes his findings about the first human electroencephalogram

1929 Karl Lashley defines equipotentiality and mass action

1928 Edgar Douglas Adrian publishes *The Basis of Sensation*

1929 Joseph Erlanger and Herbert Spencer Gasser publish work on the correlation of nerve fiber size and function

1929 Walter B. Cannon coins the term *homeostasis*

1930 John Carew Eccles shows central inhibition of flexor reflexes

1931 Ulf Svante von Euler and J. H. Gaddum discover substance P

1932 Max Knoll and Ernst Ruska invent the electron microscope

1932 Jan Friedrich Tonnies develops multichannel ink-writing EEG machine

1932 Jan Friedrich Toennies and Brian Matthews design the differential amplifier

1932 Smith, Kline, and French introduce the first amphetamine, Benzedrine

1933 Ralph Waldo Gerard describes first experimental evoked potentials

1935 Ward C. Halsted establishes the first clinical neuropsychological laboratory in the United States

1935 Dexedrine (an amphetamine) introduced to treat narcolepsy

1935 Frederic Bremer uses cerveau isole preparation to study sleep

1936 Egas Moniz publishes work on the first human frontal lobotomy

1936 Walter Freeman performs first lobotomy in the United States

1937 James Papez publishes work on limbic circuit

1937 Heinrich Kluver and Paul Bucy publish work on bilateral temporal lobectomies

1937 James W. Papez develops visceral theory of emotion

1937 John Zachary Young suggests that the squid giant axon can be used to understand nerve cells

1938 B. F. Skinner publishes *The Behavior of Organisms*, which describes operant conditioning

1938 Albert Hofmann synthesizes LSD

1938 Ugo Cerletti and Lucino Bini treat human patients with electroshock

1938 Franz Kallmann publishes *The Genetics of Schizophrenia*

1938 Ames Room designed by Adelbert Ames Jr.

1939 Carl Pfaffman describes directionally sensitive cat mechanoreceptors

1939 Nathaniel Kleitman publishes *Sleep and Wakefulness*

1942 Judith Graham develops a KCl-filled glass electrode for recording muscle fiber resting membrane potential

1942 Stephen Kuffler develops the single nerve-muscle fiber preparation

1943 John Raymond Brobeck describes hypothalamic hyperphasia

1946 Theodor Rasmussen describes the olivocochlear bundle (bundle of Rasmussen)

1946 President Harry S. Truman signs the National Mental Health Act

1947 German neurologist Joachim Bodamer coins the term *prosopagnosia* (face blindness)

1949 Kenneth Cole develops the voltage clamp

1949 Horace Winchell Magoun defines the reticular activating system

1949 John Cade discovers that lithium is an effective treatment for bipolar depression

1949 Giuseppi Moruzzi and Horace Winchell Magoun publish *Brain Stem Reticular Formation and Activation of the EEG*

1949 Donald Olding Hebb publishes *The Organization of Behavior: A Neuropsychological Theory*

1950 Karl Lashley publishes *In Search of the Engram*

1950 Eugene Roberts and J. Awapara independently identify GABA in the brain

1950 French chemist Paul Charpentier synthesizes chlorpromazine, an antipsychotic drug

1951 MAO inhibitors introduced to treat psychotics

1951 B. F. Skinner describes "shaping" in a paper titled "How to Teach Animals"

1952 The *Diagnostic and Statistic Manual of Mental Disorders* (DSM) was published by the American Psychiatric Association

1953 Brenda Milner discusses patient H.M., who suffers from memory loss of hippocampal surgery

1953 Eugene Aserinski and Nathaniel Kleitman describe rapid eye movements (REM) during sleep

1953 H. Kluver and E. Barrera introduce Luxol fast blue MBS stain

1953 Stephen Kuffler publishes work on center-surround, on-off organization of retinal ganglion cell receptive fields

1954 James Olds describes rewarding effects of hypothalamic stimulation

1954 John Lilly invents the isolation tank

1954 Chlorpromazine approved by the U.S. Food and Drug Administration

1956 L. Leksell uses ultrasound to examine the brain

1956 National Library of Medicine named (was the Army Medical Library)

1956 Rita Levi-Montalcini and Stanley Cohen isolate and purify nerve growth factor

1957 W. Penfield and T. Rasmussen devise motor and sensory homunculus

1957 The American Medical Association recognizes alcoholism as a disease

1958 Haloperidol introduced as a neuroleptic drug

1959 P. Karlson and M. Lusher coin the term *pheromone*

1960 Oleh Hornykiewicz shows that brain dopamine is lower than normal in Parkinson's disease patients

1961 Levodopa successfully treats parkinsonism

1962 Eldon Foltz performs the first cingulotomy to treat chronic pain

1965 Ronald Melzack and Patrick D. Wall publish gate control theory of pain

1965 Drug Abuse Control Act passed

1968 Alexander Romanovich Luria publishes *The Mind of a Mnemonist; A Little Book About a Vast Memory*

1969 D. V. Reynolds describes the analgesic effect of electrical stimulation of the periaqueductal gray

1969 The Society for Neuroscience is formed

1972 Jennifer LaVail and Matthew LaVail use horseradish peroxidase to study axonal transport

1972 Godfrey N. Hounsfield develops X-ray computed tomography

1973 Candace Pert and Solomon Snyder demonstrate opioid receptors in brain

1973 Sinemet is introduced as a treatment for Parkinson's disease

1973 Timothy Bliss and Terje Lomo describe long-term potentiation

1974 John Hughes and Hans Kosterlitz discover enkephalin

1974 M. E.Phelps, E. J.Hoffman, and M. M.Ter Pogossian develop first PET scanner

1974 First NMR image (of a mouse) is taken

1975 John Hughes and Hans Kosterlitz publish work on enkephalins

1976 Choh Hao Li and David Chung publish work on beta-endorphin

1976 Erwin Neher and Bert Sakmann develop the patch-clamp technique

1987 Fluoxetine introduced as treatment for depression

1992 Giacomo Rizzolatti describes mirror neurons in area F5 of the monkey premotor cortex

1993 The gene responsible for Huntington's disease is identified

2013 The start of the Human Brain Project was announced

2013 U.S. President Barack Obama announces the Brain Research through Advancing Innovative Neurotechnologies (BRAIN) Initiative

Source of timeline: adapted from http://faculty.washington.edu/chudler/hist.html.

REFERENCES

Afifi, A.K., and Bergman, R.A., *Functional Neuroanatomy* (New York: McGraw-Hill, 1998).

Albert, D.M., *Dates in Ophthalmology. A Chronological Record of Progress in Ophthalmology Over the Last Millennium* (New York: Parthenon, 2002).

Bennett, M.R., The early history of the synapse: From Plato to Sherrington, *Brain Research Bulletin*, 50:95–118, 1999.

Brazier, M.A.B., *A History of the Electrical Activity of the Brain* (London: Pitman, 1961).

Brazier, M.A.B., *A History of Neurophysiology in the 19th Century* (New York: Raven Press, 1988).

Clarke, E., and Dewhurst, K., *An Illustrated History of Brain Function* (Berkeley: University of California Press, 1972).

Clarke, E., and O'Malley, C.D., *The Human Brain and Spinal Cord* (Berkeley: University of California Press, 1968).

Finger, S., *Origins of Neuroscience* (New York: Oxford University Press, 1994).

Finger, S., *Minds Behind the Brain: A History of the Pioneers and Their Discoveries* (New York: Oxford University Press, 2000).

Francis, R.L., *The Illustrated Almanac of Science Technology and Invention* (New York: Plenum Press, 1997).

Glickstein, M., *Neuroscience. A Historical Introduction* (Cambridge, MA: MIT Press, 2014).

Gross, C.G., *Brain, Vision, Memory. Tales in the History of Neuroscience* (Cambridge, MA: MIT Press, 1998).

Harding, A.S., *Milestones in Health and Medicine* (Phoenix, AZ: Oryx Press, 2000).

Kandel, E.R., and Squire, L.R., Neuroscience: breaking down scientific barriers to the study of brain and mind, *Science*, 290:1113–20, 2000.

Marshall, L.H., and Magoun, H.W., *Discoveries in the Human Brain* (Totowa, NJ: Humana Press, 1998).

Martensen, R.L., *The Brain Takes Shape. An Early History* (New York: Oxford University Press, 2004).

Millon, T., *Masters of the Mind. Exploring the Story of Mental Illness from Ancient Times to the New Millennium* (Hoboken, NJ: John Wiley and Sons, 2004).

Pickover, C.A., *The Medical Book. From Witch Doctors to Robot Surgeons, 250 Milestones in the History of Medicine* (New York: Sterling, 2012).

Rose, F.C., and Bynum, W.F., *Historical Aspects of the Neurosciences. A Festschrift for Macdonald Critchely* (New York: Raven Press, 1982).

Sebastian, A., *Dates in Medicine. A Chronological Record of Medical Progress over Three Millennia* (New York: Parthenon, 2000).

Shepherd, G.M., *Foundations of the Neuron Doctrine* (New York: Oxford University Press, 1991).

Swartz, B.E., and Goldenshon, E.S., Timeline of the history of EEG and associated fields, *Electroencephography and Clinical Neurophysiolology*, 106:173–76, 1998.

INDEX

Note: Italicized page locators refer to illustrations; tables are noted with *t*.

ABOUT THE AUTHORS

ERIC H. CHUDLER, PHD, is a neuroscientist in the Departments of Bioengineering and Anesthesiology and Pain Medicine at the University of Washington. He is also the Executive Director of the Center for Sensorimotor Neural Engineering in Seattle. For the past 25 years, Dr. Chudler has worked to develop resources to help people learn about the brain.

LISE A. JOHNSON, PHD, is a neural engineer, science writer, and educator at the University of Washington, as well as a very busy mom to two small children. She is currently the University Education Manager and a research scientist at the Center for Sensorimotor Neural Engineering. Her research focuses on sleep, learning, and brain-computer interfaces.